BEYOND BIOLOGY

BEYOND BIOLOGY

Human evolution and the emergence of a cultural being

BY

MATTHEW B. HALLINAN

Whitworth & I
Santa Fe, New Mexico

Whitworth and i
223 N. Guadalupe Street, #283
Santa Fe, New Mexico 87501
www.whitworthandi.com

Library of Congress Control Number: 2020951846

ISBN 978-0-9771655-2-0

First published January 2021

Acknowledgements

This is an ambitious book. It was motivated by the conviction that humanity has reached a new stage in its development, and that to cope with the complex of problems we currently face, we must elevate our understanding of ourselves—what we are, how we got here, and how we fit into the natural world.

The book draws on a lifetime of research and reflection, and took over a decade to organize and write. I could never have worked my way through all of this without the help and feedback of a number of patient and generous collaborators.

In particular, I would like to thank Kathy Chetkovich for lending her remarkable writing and editing skills to the effort to make my syntax and prose readable. Dr. Allen Johnson, who I have known since our undergraduate days at UC Berkeley, spent hours over breakfasts at Saul's sharing his anthropological knowledge and helping me to probe the boundaries of my hypothesis. No one gave more time than Jack Kurzweil, who combined merciless criticism with unwavering support. Jack Radey, whose wit and wisdom—"never use a 50 cent word when a two-bit one will do"—kept me focused on who I was writing for. And a special thanks to my brothers Patrick and Conn, whose enthusiasm for the project provided an essential boost during those critical moments when it appeared that the questions I was asking may have exceeded my capacity to adequately answer them.

And lastly, I must thank my dear wife Stephanie, who besides lending her language and typing skills, tolerated—to a point—the hermit-like existence that is the lot of a writer.

"The aim of knowledge is to divest the
objective world that stands opposed to
us of its strangeness and … to find
ourselves at home in it."

G.W.F. Hegel

CONTENTS

Preface

Because this chapter comes at the beginning of the book, protocol would suggest that it be called a preface. But it's really more of a postscript. It could not have been written until the book had been finished.

Every book on human evolution begins with the assumption that we know where our evolutionary process began and we know where it led. The issue, then, is to find a hypothesis that can connect these two different points, that can explain how we got from there to here—from great ape to what we humans are today. It turns out that is not what the problem is at all.

The real problem is that we don't know where here is—what we are. We know a lot about ourselves, but we don't get the big picture. We don't know how we fit into the natural world. We can't sort out where we are the same and where we differ from other animals. We do know where our evolutionary process began--with a species of great ape living in East African some three million years ago. But it's not possible to formulate a hypothesis about human evolution without understanding where the process ends. If you don't know that, you won't be able to understand what was driving it.

What all of this means is that I could not have begun this work with a finished hypothesis. You have to get into the water before you can learn how to swim. This book, then, was my discovery process. I could not really understand what we are and how we got here until I had finished it. If I were to write it today, I would change the argument in a number of ways. But frankly, I don't have the energy for that. And I also believe there is value in taking the reader through the same learning process I went through.

Intelligence

I felt the need to write this preface because this is an unusual book. On one level, it would seem to be a typical scientific work on human evolution. I begin by putting forth a hypothesis about how our species might have evolved and then I set out to support that hypothesis with the latest discoveries from paleoanthropology. But it's not a typical scientific work. It couldn't be. Humanity is not a typical life form.

My thinking about humanity begins with difference. To understand ourselves we need to know how we became so different. This focus distinguishes this work from that of the vast majority of human evolutionists. They see evolution as proof of our continuity with the animal world. They assume that, having gone through the same Darwinian process as every other animal, humans could not be different in any fundamental way. As with the others, our behavior must be an expression of biological drives.

There is continuity between humanity and the animal world. But there is also a critical difference. And that difference can be summed up in a single word--intelligence. After writing this book, I can sympathize with those who would like to keep intelligence out of the picture.

Intelligence poses a whole set of difficult problems for evolutionary theory. To begin with, no one knows what it is. There is no consensus on how to define the term, let alone an agreed upon model of how the mind works. But the problem is even deeper than that.

Intelligence and biological evolution are opposing principles. Darwin described the process by which animal biology adapts to the rigors of some environment: polar bears evolve heavy fur and fat insulation, birds evolve hollow bones to lighten their bodies, etc. Intelligence, on the other hand, enables an animal to adapt without having to alter its biology. It allows coyotes to adapt to life in the big city.

All animals are equipped with some level of intelligence. But as smart as some of them they may be, no animal except us has an intelligence that it can make the foundation of its survival behavior. All others depend primarily on the biological equipment they are born with (claws, canines, fur coats, four-chambered stomachs, etc.). All of them adapt to changes in their environment through a Darwinian process. We don't.

Culture

While intelligence lies at the center of human evolution, it is inseparably connected to something else. If an animal's organism is the only mechanism it has to deal with the world, intelligence won't be able to carry it very far. A dumb antelope that is fast has a better chance of escaping from a predator than a smart one that is slow. When intelligence really took off is when one animal reached an intellectual level where it could figure out how to develop new capacities by using things that lay outside its organism—sticks, stones, and bones—tools. The ability to acquire new capacities by making use of the properties of external objects is the root of culture.

A growing reliance on the use of external objects shifted the whole trajectory of human evolution. It opened the path to what we are today. We did not adapt directly to some environment; we adapted to our unique way of adapting to the environment—to culture. This produced a very different kind of life form. That's the story I tell in this book.

Unchartered Territory

The problem in seeking to confirm a new hypothesis is that the bulk of existing research has been focused on questions raised by the old one.

The questions posed by this new evolutionary model have mainly to do with intelligence. To be able to make my case, I needed a general model of intelligence; one that would allow me to answer two difficult questions. First, what cognitive threshold enabled our great ape ancestors to begin to take a cultural path—to solve problems in new ways? And second, what change produced the cognitive explosion that took place around 100,000 years ago and which was associated with the appearance of contemporary *Homo sapiens*? For this, I would need to determine the way in which language affects intelligence.

The problem I faced is that there is no scientific consensus on either of these issues. There is no universally accepted model of a general intelligence, and no one really knows how language affects cognition. There are a lot of good ideas and pieces of the puzzle floating around, but if I was going to present a coherent evolutionary hypothesis, I would have to cobble together a model of intelligence on my own.

The result is that this work is something of a hybrid. Mostly it's science—built on the latest research. But it's also philosophy. That was unavoidable. In this book, I am trying to understand the emergence of a being that brought intelligence into the world as the foundation of its way of life. However, intelligence doesn't fossilize. It left some important markers along our evolutionary trail, primarily in the form of larger brain cases, improved technologies, and evidence of the emergence of a language capacity. But in order to understand humanity's intellectual progression, I had to go outside anthropology and draw on linguistics, neurophysiology, and most importantly, philosophy. The latter played a critical role in my ability to construct a framework that could make sense of human intellectual development.

I don't see this book as the final word on human evolution. My goal is to introduce a new way of looking at what we are and how we fit into the natural world. Undoubtedly, there is much that will not stand the test of time. But my hope is to turn human evolutionary science in a new and more fruitful direction.

We humans are at a critical point in our relationship with nature. We need to better understand what we are—how we got into this mess and how nature has provided us with the means to get out of it.

Chapter 1

How Did We Get into a War with Nature?

We humans are an intelligent life form. Our capacity to process information at a higher level than other animals provides us with powers that no other creature possesses. This has enabled us to overcome every obstacle to our growth process—obstacles that limit the proliferation of other species. We are able to invade inhospitable ecological zones, multiply our food resources, control the supply of water, overcome the ravages of disease, and create social institutions that enable us to construct communities that number in the hundreds of millions.

However, we currently face the most formidable obstacle of all—our own growth process. Ironically, it is our success as a species that poses the gravest danger to our continued survival. Our growing numbers and the impact of our ever-evolving technologies are taxing the carrying capacity of our planet, placing us on a collision course with the natural processes that make it habitable.

We have to turn things around. We either make the changes required to bring our developmental process into line with the laws governing the natural world, or—if past massive die-offs are a guide to the future— we could actually go extinct.

The adjustments we need to make are not easy, but they are achievable. We know what has to be done; we have the scientific and technological capabilities to stop the bleeding and set ourselves on a more nature-friendly trajectory. Our real problem is not so much with the natural world—it's with ourselves and our inability to mobilize the collective will to make the necessary changes.

A big part of the problem is ignorance. Large numbers of people lack sufficient information and, indeed, the scientific literacy to make sense of the processes that are unfolding around us. Ignorance, however, is not the only problem. An angry, over-heating planet is providing all the evidence anyone would seem to need to realize the direction we have to take. The problem is not just lack of familiarity with the science or the facts, but rather a refusal to accept them—the rejection of unwelcome information.

In the United States, the global-warming-denial movement is largely funded by the Koch brothers—who between them share three master degrees from MIT in chemical and nuclear engineering. However, they also shared, until David Koch's death, ownership in a $100 billion conglomerate that depends upon the use of fossil fuel.

When it involves billionaire corporate executives, it is easy to label such behavior as short-sighted greed. But when it involves the resistance of coal miners and lumberjacks who see their livelihoods threatened by environmental concerns—we don't have a good term for that. Perhaps the problem was best summed up by Upton Sinclair when he noted wryly that "it is difficult to convince a man of something if his salary depends on him not believing it."

Human societies are intricately organized life-support systems; every change that takes place has real-life consequences and these cannot help but filter our perceptions and influence the choices we make about which ideas to believe.

During times of crisis, humans are at their best—and at their worse. While it is imperative that the whole of humanity work together to address this planetary crisis, nevertheless, we see movement in the opposite direction. A "me first, take what you can" nationalism is gaining ground; old ethnic hostilities are being stoked; and racial prejudices, religious conflicts and divisions of all sorts are on the rise. This is a bad time for humanity to focus on its differences.

There is, thankfully, another side to what's taking place: not all is darkness. We also see the growth of a global movement that seeks to transcend historical and cultural differences—to meet the challenges we face together. Here we find a sense of universal human solidarity, mutual respect and even affection—and an embrace of the natural world. These are the qualities we will need to succeed. This emerging community embodies the possibility of humanity's future—but it is still in its infancy, struggling to get on its feet.

There it is in stark relief: animal impulse and higher purpose. These conflicting tensions have always driven the human saga. Sometimes one has carried the day; sometimes the other—but more often, some kind of mixture of both. And that is to be expected; we are a mixture of both. This is not about some eternal metaphysical drama—a battle between good and evil. This tension, rather, is a reflection of the natural process that gave birth to us— it is rooted in the kind of being we have evolved into.

A Different Kind of Being

We are, as the opening sentence declares, an intelligent being. Our successes are due to our ability to understand how things work. Little by little we have uncovered nature's principles and have used these to further our own ends. But there is a big hole in our knowledge. While we have reached a point where we can recognize the need to establish a new kind of relationship with nature, we don't really know what that entails. The truth is, we don't know what we are—we don't know how we fit into the natural world. We know where we are—but we don't know how we got here—how such a unique being could have emerged out of the animal world.

This declaration of ignorance might seem like a gross exaggeration. One of the great achievements of the second half of the twentieth century was the discovery of a more or less continuous sequence of fossils that connects contemporary humans to a lineage of great apes that lived in East Africa some 3 million years go. And we know a lot more than that. We know that those apes were members of a branch that broke off from the reptilian main stream some 250 million years ago (mammals). And we can go further than that—we can trace those early mammals all the way back to life forms that emerged from the substance of our planet some 3.7 billion years ago.

We know all of this—but we don't know what it means: what it implies about the kind of being we are.

Until the 1970s, anthropology and the other social sciences did not see evolution as relevant to an understanding of ourselves—to how we behave; to the substance of our goals and dreams. Evolution is about biology. Natural selection, Darwin's great discovery, describes the process by which animal populations adapt to their environments. No one questions the validity of that discovery, but it is not at all clear what it has to do with us.

Animals adapt to their environment with their biology. Polar bears have evolved thick fur and layers of fat to insulate them from the frigid Arctic climate. Inuit people, on the other hand, make heavy layered clothing, seal-skin boots, and build igloos which they warm with oil lamps. Humans don't adapt to their environment with their biology. They invent technologies, shelters, clothing, organizational forms, conceptual formulae, etc. They use their intelligence to create devices that give them capacities that are not built into their biology. With humans there is a distinction between biology and behavior that exists nowhere else in the animal world. Other animals depend on the capacities they are born with, and those capacities are inseparably connected to their physical make-up. Humans, on the other hand, depend upon capacities we acquire—the knowledge, lore, technologies and social connections we have access to as members of a given historical community. Anthropology called this storehouse of accumulated know-how and devices culture. Humans relate to their environment by means of their culture, not their biology. And culture is the product of invention; it develops by means of its own accumulative processes, not by evolutionary ones.

This distinction between biology and culture was anthropology's great contribution. Anthropology emerged as an academic field in the late nineteenth century, a time when a number of European powers were jostling with one another in a race to expand the reach of their colonial empires. Darwin's ideas were being used to support the notion that higher levels of industrial and military development were the work of more highly evolved races. These races, it was argued, had the right, indeed, the duty to rule over their less developed brethren; "survival of the fittest" was conflated with "might makes right."

The idea of culture played an important role in combating the view that historical differences were expressions of different stages of evolutionary development. Anthropologists were in the forefront of opposition to the various racial theories and ethnic blinders that would play such a big part in leading Europe into two horrendous wars. There is something clearly right about anthropology's notion of culture: but there is also something missing in it. The idea of a complete separation between human biology and human behavior is not credible; it clashes with everyday experience. Humans have powerful inner strivings and needs; they are not merely the products but also the creators of their cultures—and they change them when they fail to meet their needs. Anthropology's notion of humans as the passive products of cultural processes—the "clay," in the words of Ruth Benedict, out of which the "cultural potter" fashions its human artifacts—could not account for social turmoil, much less, revolutionary change. And, indeed, it began to fall into disfavor during the tumultuous events of the 1960s.

Sociobiology

During the 1960s and 1970s, new ideas about the relevance of biology and evolution began to stir in the academic community. This thinking first appeared among evolutionary anthropologists, and then, with the publication of E.O. Wilson's *Sociobiology* (1975), the broader biological community was brought into the discussion.

Anthropology had built its case on the perception of a fundamental distinction between human and animal behavior. Animal behavior was built into their biology, while human behavior was learned—acquired from one's culture. This distinction created a gap between us and the animal world. This new school rejected the notion of a gap, arguing that such a discontinuity was not possible—that it violated the laws of evolutionary biology.

Humans, it was argued, are first and foremost biological beings. As such, we could only have got to where we are through an evolutionary process guided by natural selection—the same as every other animal. That fact, the evolutionists argued, has certain inescapable implications.

Natural selection was Darwin's great discovery. The basic idea is quite simple. Any change that helps an animal to better survive in a given environment will enable that animal to leave more offspring than those members of its species that lack that feature. Over time, as a result of this reproductive differential, the entire population will come to consist of individuals that possess the beneficial change. That, according to Darwin, is how evolution takes place.

The implication of this is that every aspect of an animal's biology is the product of a "selection" process; everything can be traced to some contribution it made to that animal's evolutionary success. This process produces creatures that are biologically adapted to their environment and that are equipped to pursue more or less effective survival strategies. Every biological creature, including us, is the product of this evolutionary dynamic. Every step in the path by which our great ape ancestors evolved into human beings was guided by natural selection.

If everything about us exists because of some evolutionary advantage it conferred, it would not seem possible that our behavior could be divorced from evolutionary goals. Our behavior, whether we are conscious of it or not, is directed by the same kind of self-seeking ends that motivate other animals.

What about higher ideals—morality and altruistic behavior? These cannot possibly be explained in terms of some biological or evolutionary advantage. Yes they can, E.O. Wilson assures us. Humans like to believe they are selflessly pursuing higher ideals, but according to him, everything we do is based upon an underlying, "hard wired," and largely unconscious evolutionary calculus. It all comes down to maximizing one's genetic inheritance—leaving more offspring than our competitors. If a creature is not acting according to that logic, it will soon be replaced by one who is.

What about culture—the distinction that anthropology draws between behavior that is learned rather than built into an organism's biology? There is nothing, Wilson argues, that is uniquely human about culture. Culture is simply a form of "tradition:" learned behavior that is passed down from one generation to the next. Such behavior is widespread in the animal world; elephant herds, orca pods, and chimpanzee troops all serve as repositories of accumulated experience; they all have their own distinctive "cultures." While humans may be more adept at transmitting learned behavior than other animals, the difference is only relative. Culture, Wilson tells us, "differs from animal tradition only by degree"(Wilson 1975, p. 168).

Once culture has been redefined as a form of animal behavior, there is nothing left that fundamentally distinguishes humans from other animals. There is no gap; nor could there be. Humans are on a continuum with the animal world. We are different, but no more different from them than they are from one another.

Like anthropology, sociobiology is based on something real. It brought the animal in us back into the picture and it laid down an important principle: humans are the products of an evolutionary process and any model that attempts to understand what we have become must be consistent with the laws of evolutionary biology.

The gap that anthropology believed separated humans from other animals—the distance it perceived between biology and culture—was too great. There was no evolutionary path that could lead from the animal world to the model of humanity that anthropology had constructed.

Sociobiology represented something of a corrective to anthropology's cultural one-sidedness, but it went too far in the other direction. While confidently asserting that everything that humans do could be explained in terms of evolutionary biology, in fact, sociobiology never truly addressed the issues that were central to anthropology's case. Animals adapt to nature with their biology; humans adapt with their culture. No animal's relationship to nature is determined by its culture; no human's relationship is determined by its biology. Sociobiology danced around these facts without ever directly confronting them.

Humans can alter their ways of life, the way they relate to nature and one another, without altering their biology. There simply is no way to force this reality into the same mold we use to make sense of animal behavior.

Unforeseen Consequences

In claiming that there was nothing truly distinctive about human behavior, that culture was just a form of animal "tradition," sociobiology eliminated the core idea in anthropology's separation of biology from behavior. By so doing, it removed the firewall that anthropology had built against biological and racial interpretations of human historical differences. If behavior is an expression of biology, then it would seem to follow that different levels of technological and organizational attainment should be the products of different biologies.

This is not how Wilson, or the great majority of evolutionary anthropologists, intended their ideas to be read. Indeed, Wilson explicitly rejected the idea that historical differences between humans were the products of different, built-in biological capacities. He argued that differences between human communities were actually very small, and that the entire range of human variability could be generated by a single genome. That argument, however, does not hold up.

Different societies are not that similar. The skill set of a computer programmer living in a twenty-first century urban setting is vastly different from that of a hunter/forager living in the Amazon forest. Cultures are very different, and require very different kinds of technological, social and cognitive aptitudes. Indeed, it is not difficult to come up with all kinds of hypotheses about how certain cultures might select for greater intelligence or superior moral or martial qualities.

While the racial implications of sociobiology's biological framework were largely ignored when those ideas were first formulated, that's not true today. One only has to go on to white nationalist websites to find a wide range of "scientific" anti-Semitic and racial theories—all based on the idea that biology and behavior are inseparably connected.

The discussion we are having here is not some abstract academic exercise. Getting the relationship of human biology and behavior right is a necessary part of humanity coming to understand its oneness.

Getting Beyond the Differences

Much has happened since 1975 when Wilson shook up anthropology's idealized cultural world. Indeed, Wilson himself has gone through a whole cycle of changes. However, the basic opposition between a cultural and a biological perspective has not been bridged; it still largely remains in force.

Cultural anthropology has pretty much gone its own way, having concluded that evolutionary issues are not relevant and that they simply open up negative debates. E.O. Wilson's new ideas, which we will discuss later, have created turmoil among his followers—but he still holds to the idea that human behavior is an expression of evolutionary processes. He just has a different idea about what those processes were.

The most interesting development has been the emergence of a current that seeks to find some middle ground. Basically, they build on Wilson's idea that culture is tradition—that it consists in the transmission of learned behavior.

Following Michael Tomasello (1999), this current sees human evolution as taking shape around an increased capacity to share and accumulate information. There is value in this perspective; it is a part of the equation. But it does not give us the big picture. Human culture is not just animal tradition writ large, and human evolution is not simply an extension of the same evolutionary processes that operate elsewhere in the animal world.

We are really different—and to understand what we are, we need to get past certain conventional ways of looking at things and develop some new analytical tools.

Rethinking the Whole Process

Sociobiology's great strength is its starting point; it begins by placing the discussion within an evolutionary context. The one thing we know for sure is that humans could only have got here through an evolutionary process. This idea, however, gets conflated with a different one—that there is only one kind of evolutionary process. If this were the case, that we were the products of the same kind of process as every other animal, then it would stand to reason that our behavior should embody the same evolutionary logic as theirs.

But is this true? Is there only one kind of evolutionary process? This idea has been a major road block to the development of new ways of thinking about human evolution. What has made it so difficult to get past is that it is not all wrong. On one level, there is something common to every evolutionary process: the mechanics of natural selection. Darwin got that right. However, the fact that humans and animals are equally the products of natural selection doesn't mean that we must therefore be fundamentally the same as them. Natural selection is not some kind of levelling process. In fact, it produces variety; it is the mechanism that drives diversity. It has produced all the different kinds of life forms—from bacteria, to jellyfish, elephants and *Tyrannosaurus rex*. While every evolutionary process shares something in common, at the same time, each is different from every other.

The question of whether humans can be a different kind of creature has nothing to do with whether our process was guided by natural selection. The real question is: What kind of natural selection? Every animal is the product of a distinctive form of natural selection. Could there have been something unique about the selective process that shaped our evolutionary development? The question is this: Is it possible to imagine a selective process that was so different that it resulted in a unique kind of being, one that could not be categorized as simply another kind of animal?

That's the issue. And it raises the question: What were the selective pressures that guided human evolution? Natural selection is about adapting to some niche—to some way of surviving in a given environment. Both lions and cheetahs have adapted to the African savanna; but one preys upon large, powerful herbivores, and the other on smaller, fleeter ones. Each then has a different niche. What was the niche our ancestors were adapting to? This poses a really difficult and almost universally avoided question: How do we define the human way of life?

The Human Way of Life

To understand the forces that shaped human evolution, we need to determine what our ancestors were doing to survive. The early members of our lineage distinguished themselves from their great ape ancestry in three major ways. First, they abandoned the trees, and their bodies adapted to life on the ground. Second, they possessed exceptionally large brains—the earliest members of our lineage had brains that were 35% larger than those of their australopith ancestors—who already had the largest brains in the animal world. Our lineage went on to evolve brains that were three times the size of the australopiths and seven times that of the average mammal. And lastly, while our first, human-like ancestors adopted meat-eating as an important part of their diet, they did not evolve protruding canines or claws. They had no biological means for killing prey, nor for that matter, defending themselves. They relied on devices they constructed—weapons and tools.

These changes represented a radical evolutionary turn-about. An arboreal, vegetarian great ape abandoned the trees, came down to the ground and became a predator—a meat eater. It did not evolve the normal biological accoutrements of a predator, but relied instead on tools and weapons it manufactured. Lastly, and most importantly, it experienced an unparalleled expansion of its brain. What was going on here? How were our ancestors adapting to life on the floor of the savanna? What kind of natural selection was shaping their extraordinary evolutionary process?

Based on what we know about these early humans, and drawing on our ability to see where their development was heading, we can assume that their survival depended primarily on three behavioral capacities: the ability to make and use tools; the ability to communicate with signs and symbols (some early form of language);

and organization—the ability to coordinate their actions. These three behaviors provided our clawless and canine-less ancestors with the means to move up the food chain. These behaviors create a power gap that gives humans an advantage over the other creatures on this planet.

But how unique are these behaviors; are they truly outside the range of normal animal behavior?

What are These Behaviors?

What is so distinctive about technology, language, and organization? While each of these behaviors are different from one another, at the same time, they share something important in common. All of them make use of something outside a creature's body that gives it a capacity that is not built into its physical organism. With technology a being makes use of the properties of external natural objects. Our early ancestors didn't need to evolve canines and claws when they could use digging sticks and stone tools to gain access to the foodstuffs on the floor of the savanna. With language we use symbols—external markers—that enable us to represent and share our thoughts. Our early ancestors used gestures and signs. This allowed them to collectively process and exchange information. And lastly, as a result of our unique communication capacities—our ability to share goals and plans—we are able to make use of the energies of other humans in unprecedented ways. The ability to use signs to coordinate behavior enabled our early ancestors to combine their forces into social "tools" that produced force levels that only the most powerful animals could challenge.

These three behaviors tool using, communication with signs, and organization—are ways an animal can acquire new capacities without changing its biology. Looked at from an evolutionary perspective, this is the essence of the difference between the way that humans and animals survive. What distinguishes us is that we don't adapt to nature with our biology. We adapt with technology, language (our capacity to share our thinking and knowledge), and the ability to organize and combine our energies. This combination of behaviors forms the core of culture. Humans do not survive on the basis of their biological resources alone; we survive by making use of the properties of things that are not part of our organisms.

This way of life completely alters the way natural selection works. It is based on an indirect relationship with nature. Our ancestors were no longer adapting to their environment with their physical organisms; their survival depended, instead, on their ability to create and make use of technological, communicative, and social "tools." Under these circumstances, natural selection switched, and instead of shaping our ancestor's biology to directly cope with the rigors of some environment, it adapted their biology to better perform these new kinds of behavior. Our species, then, is the product of a unique kind of natural selection. We adapted to culture rather than to nature.

A New Kind of Being

The perspective presented here provides a basis for formulating a new hypothesis about human evolution. Our species is the product of a different kind of evolutionary process: we adapted to culture rather than directly to nature.

But what does that mean? That's what this book is about.

Our evolutionary process did not give rise to another kind of animal; it gave birth to a new kind of life form. Our biology evolved as a part of a life process that extends beyond our organisms and that incorporates forces that are not of a biological nature and that give us capacities that enable us to transcend the limits of biological form. That's the story I am going to tell in the following pages. It gives us the best fit with the facts—but most importantly, it solves what has always been thought of as two contradictory goals. It connects us to the animal world—finds an evolutionary path that shows how a great ape could have evolved, step by step, into us. At the same time, by showing how things that were outside our biology were incorporated into our evolutionary process, it enables us to understand how a continuous process could have given rise to a discontinuous result—to a gap that distinguishes us from all other life-forms on our planet.

Chapter 2

How Did it All Get Started?

Human evolution represented a radical departure from normal evolutionary processes. The question, then, is what could have enabled the evolutionary trajectory of one animal to radically diverge from that of every other? The critical fact is that this animal was a great ape. But even more than that, it was a particular kind of great ape. It alone possessed the prerequisites that would allow it to make this transitions.

Hands

The great apes are marvelous creatures. They are the largest animals ever to have lived in trees. For animals of their size to be at home above the ground they had to carry general primate capacities to a higher level. They have an incredible visual system: excellent color-vision and a specialized section in the retina (fovea) that allows them to perceive intricate details. These are essential factors for accurately determining the state of the branches that must bear their heavy weight. While they share many features of their visual system with monkeys, the eyes and brains of apes are uniquely wired to give them the most highly developed 3D vision in the animal world; accurately judging distance is another must for their way of life.

The way apes move—how they get from one place to another—relies to an unprecedented degree on their upper body, particularly their arms and hands. When they first broke off from the ancestral monkey lineage, sometime around 25 million years ago, they were smaller and, like monkeys, travelled on all four limbs on the top of branches. However, as they got larger, it became important for them to be able to move below the branches. For a heavy arboreal animal, hanging from the less sturdy branches is a much safer and more efficient means of movement than trying to stay on top of them. It is also an essential means of reaching fruit, which generally clusters at the tips of branches. And fruit eating, it appears, played an important role throughout their history. Over time, then, most of them evolved an anatomy that was geared for hanging and swinging, which depended on incredibly strong and very flexible arms and hands.

Their "ball and socket" shoulder joint, essential for hanging and swinging, permits an exceptional range of arm movements. Their wrist joint, unique among the primates, is built for flexibility—as is their elbow joint. Together, these two joints allow them to turn their hands 180 degrees as they swing from one arm to the other. (This is the same feature allows us humans to play on "monkey bars" and unscrew the top off a jar.)

Along with almost all other primates, the great apes have lost their primordial mammalian claws and acquired fingernails in their place. This enables them to grasp branches rather than climb with their claws, as bears do (a method that would confine them to trunks and the largest limbs). Another advantage of losing claws is that the tips of the fingers can rest directly on the surface of objects. Packed with nerve receptors, their fingertips serve as major sources of tactile information. Apes don't have to rely on whiskers, lips, and tongues to feel things the way animals with claws do.

When we put all this together, one fact in particular jumps out at us: the great apes have the most effective hands—means of manipulating objects—in the animal world. This is not due simply to the particular shape of the hands themselves, but to the whole package of features—wrist, elbow, and shoulder joint, as well as a whole complex of neurological developments that enable unprecedented hand-eye coordination. This whole suite of features played a vital role in opening up new behavioral options for our ancestors.

Our ancestors, however, were not just great apes; they were a specific branch of the great ape family—the Australopithecines. While they shared the features discussed above with the other apes, they were, at the same time, importantly different. Only the australopiths were in a position to take the critical steps that would open the door to a new evolutionary path.

Location, Location

There were two important features that set the australopiths apart from the other great apes: one was where they lived. While there is some evidence that this lineage's point of origin may have been located more towards the center of the African Continent, they are known to us almost entirely from eastern and southern Africa. This was the most geologically unstable and climatically turbulent region in the whole of Africa.

East Africa, in particular, has long been a center of geological upheaval and climate change. A major section of this region, the Horn of Africa, is slowly pulling away from the rest of the continent, creating a gigantic fissure in the earth's crust—the Great Rift Valley—that extends from the Red Sea in the north to Lake Malawi in the south. Huge upwellings of magma have elevated large portions of the region and have produced some of the world's great volcanoes. As a result of the mountain building and higher elevation, this area has a cooler and drier climate than the rest of equatorial Africa.

The tropical rain forests that once blanketed the whole continent have been receding in parts of East Africa for 20 or more million years. As the forests retreated, they were replaced by less densely wooded areas and grassy savanna. While the overall trend was towards a drier climate, the region also periodically experienced episodes of intensely wet weather. Long-term drying, yes—but the primary feature of the region was great climate and ecological volatility.

Around 3 million years ago, the climate in eastern and southern Africa entered a particularly severe phase. This appears to have been triggered by the freezing over of the Arctic Ocean and the emergence of massive ice sheets in the Northern Hemisphere. The current view is that the formation of the polar ice cap, by tying up immense amounts of water and reflecting the sun's heat back into space, produced a drier and cooler global climate. Whatever the situation globally, this was definitely the case in East Africa. The rain forests took another big hit, and woodlands, savannah, and desert further advanced. There was a major extinction event throughout the region, lasting from 2.9 to 2.6 million years ago (deMenocal, *Science*, 2011).

The dramatic shift in weather appears to have presented a serious crisis for the australopiths. They would need to adapt to these changes, or go extinct.

Bipedalism

While where they lived would require them to make changes, it was the other feature that distinguished them that determined the extraordinary direction those changes would take. They were not just great apes; they were members of a specific lineage that had followed a somewhat different path than their gorilla and chimpanzee cousins. While they shared much in common with the other two branches, they also possessed characteristics that set them apart. The distinctive evolutionary history of the australopiths and the role it played in preparing them to move in a new evolutionary direction has only recently come to light.

Beginning with Darwin's own prescient observations, the mainstream of evolutionary thought has long assumed that humans evolved from African great apes. The chimpanzee, being the ape that most closely resembles us, generally provided the model for how people envisioned our great ape ancestor. And, as DNA testing has proven, chimps are in fact our closest living relatives. Their anatomy and biological systems are strikingly similar to our own.

Nevertheless, chimps are also very different from us. So different, in fact, that it would have been very unlikely that they could ever have evolved into something like us. Their evolution took them down a path that would have made a key feature of human evolution very improbable—walking on two legs.

The anatomy of chimpanzees took shape around a particular form of arboreal travel. All great apes, because of their large size, have to be able to move under the branches. Chimpanzees, however, have evolved a "suspensionary" anatomy—a set of specialized features that are specifically designed to allow them to hang from their arms and to rely on them when traveling in the trees. They have very long arms and, most significantly, a large hand with a short thumb and long, curved fingers. The hand serves as a powerful hook for hanging and climbing.

As different as their upper bodies are, the area of greatest contrast with humans is the lower body. The lower back is short, inflexible and tightly connected to the pelvis. There is no spinal curvature. This design is helpful for stabilizing the lower body as the animal moves about with its arms. The musculature of the pelvis and legs are also very different from ours. They are constructed to enable the legs and feet to grasp and climb and are very different from what we find in humans.

While chimps do spend a lot of time on the ground, their anatomy makes standing erect and walking on two legs very awkward and inefficient. The key problem in bipedalism is holding the upper body erect while all the weight is transferred from one leg to another. Chimps are not built to do this. When they attempt to move on two legs, they have to shuffle—shift their bodies from side to side. When they do need to move on the ground, their preferred method is "knuckle-walking." They travel on all fours, relying on specially adapted features in their wrist and finger joints to enable them to use their long and powerful arms to propel themselves forward. While it is not entirely inefficient, this form of movement fails to accomplish a key prerequisite for human evolution: the freeing of the hands and arms from locomotor activity.

The australopiths, on the other hand, have a very different evolutionary history. The current best estimate of when their ancestral line—our hominid lineage—broke away from the chimp line is somewhere around 7 million years ago. Recent discoveries have taken our knowledge of this lineage back some 4.4 million years to an ancestor of the australopiths called *Ardipithecus ramidus*. This four-foot-tall, 110-pound creature has changed our thinking about the nature of our ancestral lineage.

Ardipithecus (male)

Ardipithecus was very different from contemporary chimpanzees. It had not evolved the chimp's "suspensionary" form of locomotion. It could move under the branches, but its hands and upper body were not specialized for hanging or traveling primarily with its arms. It had grasping feet, and the assumption is that it moved slowly, using all its limbs. The key differences from the chimp, however, were in the lower body. *Ardipithecus* had a longer lower back, the vertebral column was more flexibly connected to the pelvis, and there is evidence that it possessed an s-curve. The pelvis itself was a mosaic of different features—some designed for arboreal activity, and others for enabling the upper body to hold itself erect when the animal traveled on the ground. The upper leg bone—the femur—provides evidence that the gluteus maximus, a muscle that is important in holding the body erect during bipedalism, was attached more like it is in humans than in apes (Lovejoy 2009, p. 71e5).

Ardipithecus, then, had not developed the

Ardipithecus (female)

highly specialized arboreal adaptations of the chimp lineage. Its teeth were also very different. They were not specialized for fruit and soft shoots, as those of chimps are. Rather, the molars and shape of the jaw suggest a more varied diet, some of which undoubtedly was being found on the ground in the spaces opening up between the trees in its thinning forest habitat.

As the forests in East Africa continued to decline, the descendants of *Ardipithecus* would have been spending an increasing amount of time on the ground, and given the way the *Ardipithecus* anatomy had evolved, knuckle walking would not have been an option for them. They would have built, instead, upon the incipient bipedalism of *Ardipithecus.*

Sometime around 4 million years ago, the *Ardipithecus* lineage gave rise to the australopiths. Most of what we know about the early australopiths comes from "Lucy," a relatively complete skeleton of an australopith female that lived around 3.2 million years ago. Lucy's remains make clear that the australopiths had "evolved virtually all of the fundamental adaptations to bipedality" (ibid., p. 71).

In sum, the lower back, pelvis, femur, and knee joint had all been re-shaped for bipedalism. The most striking changes, however, were in the foot. The australopiths had lost the grasping foot of *Ardipithecus.* The big toe had moved in and lined up alongside the others toes, and the foot had a fully developed arch and a re-sculpted heel. All of these features are critical: they transform the foot into a platform for propelling the body forward, while at the same time cushioning the shock when the animal lands with all its weight on one foot or the other.

The lower bodies of the australopiths had gone through a major transformation and were no longer designed for life in the trees. They had been re-shaped by the biomechanical

Australopithecus

demands of terrestrial bipedalism. While the process may have begun with *Ardipithecus*, the australopiths had carried out a monumental switchover. From the waist down, they had become biologically committed to life on the ground.

This commitment is also reflected in their teeth. They had evolved larger molars with thicker enamel, indicating that they were consuming coarser, more fibrous foods: tubers, corms, and perhaps grass seeds. Microscopic studies of the teeth reveal a pitting that would seem to confirm this picture. The australopiths appear to have been adapting to life on the margins of the forest, where woodlands abutted the expanding open savannah.

The australopiths were really halfway creatures. From the waist down, they were built for life on the ground; however, they were not yet ready to completely abandon the trees. Their upper bodies—their hands, arms, shoulders, and torsos—were still designed primarily for climbing and moving about arboreally. With their in-between anatomy, the australopiths were hedging their bets. The world's most fearsome predators occupied the floor of the African savannah. Australopith males were at most five feet tall and weighed about 115 pounds. Without claws or large, powerful canines, they were essentially defenseless. The ability to climb and nest in trees would have been essential for their survival.

Tool Use

The fact that the australopiths were bipedal—that their hands were free when they moved on the ground—suggests they were particularly well suited for using tools. This whole issue—the role of tool using in the lives of the australopiths—is a focus of current research and has long been a source of controversy.

The australopiths present paleoanthropologists with a number of puzzles. Their lower bodies are built for terrestrial locomotion while their upper bodies are designed for climbing. Their skulls are shaped by the demands of erect carriage—bipedalism. The spinal cord enters the skull more from underneath than from behind as it does with the other great apes. While this serves to better balance the head for standing erect, it has an important drawback.

The main means of defense and aggression among great apes

is the use of their considerable canine teeth. The power of the canines comes from what is known as the "canine complex"—the way the jaws, skull, and powerful shoulder muscles are all connected. A chimp or gorilla bites with its whole upper body. The placement of the head in the australopiths largely separates the skull from the shoulders—essentially dismantling the canine complex. As a result, their canines could not have been truly effective weapons—and, in fact, they are very small and hardly protrude above the level of the other teeth. They clearly were not important factors in these animals' defense strategy. This is puzzling. As the australopiths were adapting to spending more time on the ground—something that would have made them increasingly vulnerable to predators—they were simultaneously losing their main biological means of defense.

Then there are the problems posed by the different kinds of foodstuffs that are found at ground level. As the forests thinned out and the richer, more tropical vegetation disappeared, the australopiths would increasingly have been forced to come to the ground in search of new food sources. These would have consisted largely of corms, tubers, and bulbs—edibles that have to be dug out of hard, sunbaked soils. Without some kind of digging implement, this would have posed major problems for the clawless australopiths.

Could they have been successful in this ape-unfriendly environment without tools?

Evidence for Tool Use in Australopiths

The use of tools by the australopiths appeared to be one of those problems that would always remain in the sphere of speculation. The living great apes, particularly chimpanzees, demonstrate an ability to use simple tools. However, experiments have shown that they are not capable of making stone tools. It was generally assumed that the australopiths, with a brain that is only marginally larger than a chimp's, would show the same limitations. That meant that the only tools they could have used would have been constructed out of organic materials—wood, bone, and horn—and these would not have been preserved over millions of years.

Then in early 2015, an ingenious study (Skinner et al. 2015) was made of the hand bones of an *Australopithecus africanus. A.*

africanus is a variety of australopith that came after the "Lucy" type (*A. afarensis*) and that lived between 3 and 2 million years ago in southern Africa. The authors of the study made use of the fact that the interior of bones—the spongy, softer material—is shaped by "mechanical loading," the stresses it is subjected to during the life of the animal. The state of this tissue can tell us a lot about what the animal was doing with its hands.

The question the authors of the study asked was whether the australopiths were capable of a "power grip"—the ability to forcefully oppose the thumb to the fourth finger. This is the key to human manual dexterity. The ape hand is not constructed to effectively perform this kind of grip; their thumbs are generally too short, not powerful enough, and placed too far away from the fingers. At the same time, the fingers of great apes are long and curved, and the power of the hand comes not from an opposition between the fingers and the thumb, but from the fingers and the palm wrapping around some object—forming a powerful "hook" that can lift or hold the animal aloft.

The study determined that the *A. africanus* hand possessed a mosaic of different features. It was more like ours than those of contemporary apes. It had a larger thumb, but at the same time it still had long, curved fingers and was designed primarily for climbing and hanging. It could not produce a fully developed, human-type precision or power grip. Nevertheless, using the latest fMRI-imaging technology, the authors of the study were able to determine that *A. africanus* had used its hands to grip objects more or less the same way humans do.

> These results show that *A. africanus* was capable of habitual and forceful human-like opposition of the thumb and fingers during precision and power (squeeze) grips that are used during tool-related behaviors, providing morphological evidence of committed tool use in a hominin hitherto considered not to be capable of these behaviors (ibid., p. 399).

For the first time, we have evidence that the australopiths were almost certainly habitual tool users. Even if the tools themselves did not survive, the fossil evidence strongly points in that direction.

Then, later in 2015, stone tools dating from 3.3 million years

Human hand *Australopithecus* hand bones

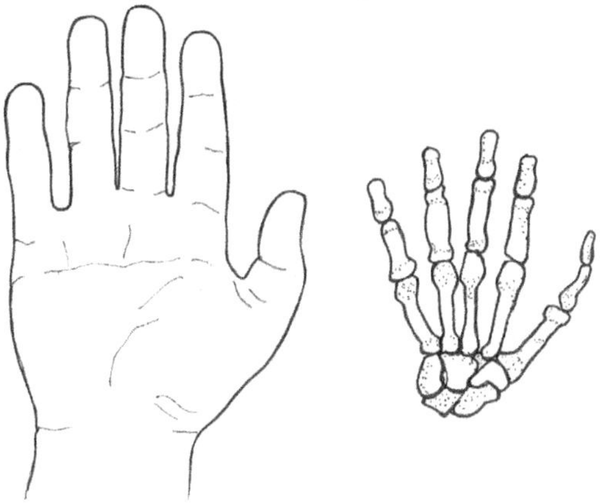

ago (*Nature* May 21, 2015) were discovered in West Turkana, Kenya. These tools were very simple and primitive, but they were beyond anything that current great apes can produce. The early date argues that the maker of these tools almost certainly was some kind of australopith. This came as quite a surprise. Australopiths were generally believed to operate within the same behavioral constraints as contemporary chimps, and this discovery suggests that this was not the case.

As long as the australopiths depended on access to the trees, they would be unable to evolve an anatomy that would permit truly efficient tool making and use. However, it is becoming clear that they were habitually using tools, and that some of them, at least, were more skilled at this than previously thought.

The Appearance of Something New

About two and a half million years ago, we begin to see evidence for the widespread use of stone tools. Before the Kenya discovery, the earliest finds had been dated to 2.6 million years. They were better made than the Kenyan tools, but at the same time, they were still quite primitive. They consisted of sharp-edged flakes produced by knocking chips off of hard, round river pebbles. These have been found in the fossilized carcasses of large herbivores—hippos, elephants, and antelope. The bones show furrows and cut marks where the sharp edges of flakes had been used to remove meat.

As simple and primitive as these flakes were, experiments have demonstrated that they were quite effective. Nicholas Toth and Kathy Schick, anthropologists who have specialized in studying early human stone technology, discovered that these simple flakes were more efficient than large canines in cutting through the thick hides of elephants and hippos. While the makers of these flakes could not have hoped to kill one of these huge animals, this stone technology would have enabled them to scavenge their corpses. Indeed, according to Toth and Schick (1993), stone flakes would have allowed their users to get the jump on their carnivorous competitors. Lions and hyenas must wait up to three days for the tough hides of pachyderms to soften before they can tear through them to reach the meat. Tool-wielding hominids, on the other hand, would have immediate access to these rich sources of protein.

No fossil remains of an australopith or other human ancestor have been found in direct association with these early stone tools. However, a number of discoveries have been made of australopith-type creatures that date from more or less the same time period. They have been given the name *Homo habilis*—"handy man." They are the first hominids to be placed in the genus *Homo*—in recognition of their assumed tool-making skills.

Homo habilis looks very much like its australopith ancestor, except for one critical feature: they have significantly larger brains. While the brains of the australopiths were slightly larger than those of contemporary chimpanzees, 425-450cc compared to 400-425cc, these new creatures had brains that varied between 550cc and 800cc. That is a huge leap. The australopiths had the largest brain in proportion to body size of any animal that had ever lived up to that point. And these new creatures had brains that averaged about 35% larger than theirs

Something extraordinary was taking place in the evolutionary history of our lineage. Big-brained ancestors that manufactured stone tools and ate meat had come on the scene. A new way of life was taking shape. The combination of effective stone tools and meat eating appears to be some kind of marker—its presence is associated with a leap in brain size and with a shift in the evolutionary direction of our ancestral line. It would seem to signal the appearance of some kind of new behavioral threshold.

Any hypothesis on human evolution must come to terms with tool using. What is it about tool using that makes it so rare—that places it beyond the capacities of the vast majority of animals? What kind of special biological platform does it require? And what kind of selective pressures does it place on an animal? Those questions must be the starting point of any investigation of the forces that shaped human evolution.

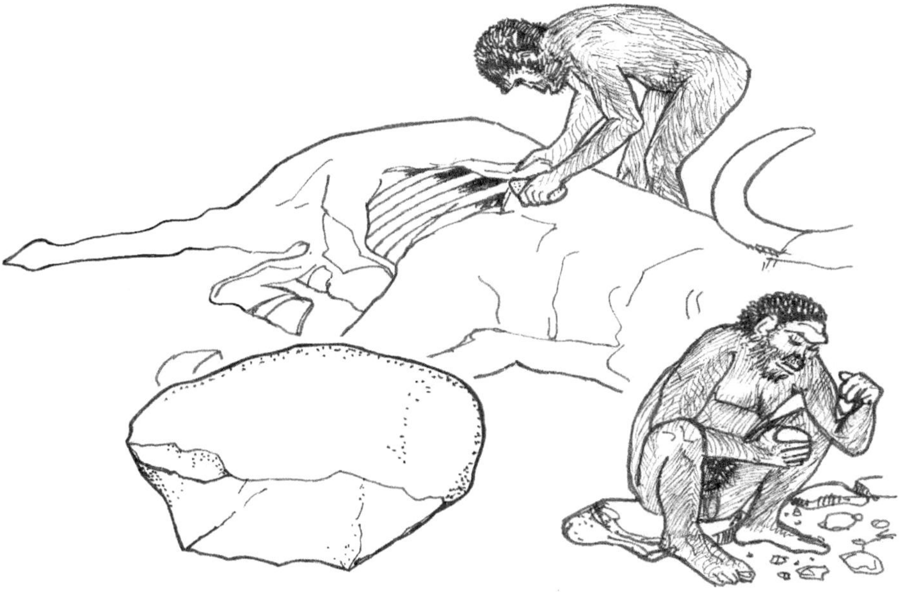

Homo habilis

Chapter 3

Tool Using and Intelligence

When animals are placed into experimental situations to test their tool-using capacities, the first thing that becomes clear is how complicated this behavior is. Because we have evolved to use tools and are taught from infancy to use all kinds of devices, it is hard for us to understand how alien tool using is to the behavioral strategies of most animals. Animals are designed to solve problems with their bodies; that's what evolution has equipped them to do. They have evolved the organic "tools" they need to survive: teeth with which to cut, chew, kill or defend themselves; claws to tear; hooves to run fast or to deliver a blow; tongues to reach distant leaves; guts to process particular kinds of food, etc. An animal's entire behavioral repertoire is built around using its organism as the means to solve its survival problems.

Tool using becomes a factor when an animal is faced with a problem it is not biologically equipped to solve. Experiments show that when most animals are put in a situation in which they cannot reach a goal without using a tool, even though a tool that would enable them to solve the problem is placed in full view, they fail to use it. Instead, they run through their biologically given behavioral repertoire—clawing, biting, pushing, digging, etc.—and when that is exhausted, they abandon the goal.

Part of the difficulty is that very few animals have an anatomy that lends itself to effective tool use. Carnivores have appendages that end in claws, while ungulates have hooves; all of which makes grasping and fine manipulations extremely difficult. Experiments have shown that crows and their relatives (New Caledonian jays, ravens, etc.) have exceptional intelligence and are good at solving certain kinds of tool-using problems. However, the fact that they have wings and must rely on their beaks seriously limits their effectiveness as tool users.

The incredibly dexterous hands and flexible, mobile arms of the great apes, products of selection for life in the trees, equip them with the best limbs in the animal world for using tools.

The Main Problem Is Intellectual

Anatomy, however, is not in most cases the main problem. The principal difficulty with tool using is that it requires a different way of looking at a problem. The vast majority of animals are equipped with an intelligence that enables them to use their organically based capacities to solve their problems. To imagine that something outside one's own organism could be brought to bear on a problem necessitates a major reorientation: the animal must cease looking at a problem in terms of its own capacities, and come to understand it as a physical puzzle in its own right. It must be able to view it in terms of cause and effect. That is critical if it is to imagine how something other than itself could influence the situation it faces. Only a handful of species are able to do this.

To clarify this issue, I would like to compare tool-using experiments carried out on two different primate relatives of ours: the cebus monkey and the chimpanzee. Both are intelligent animals and able to use tools, yet their tool-using capacities differ dramatically. The cebus cannot grasp cause and effect, and must therefore resort to trial-and-error techniques, while the chimp can understand causal relations and is able to solve the problem in its head before it acts.

The *Cebus*

The *Cebus*, popularly known as the capuchin monkey, is a New World monkey, inhabiting Central and South America. It is perhaps the most intelligent of all the monkeys, with the sixth largest brain-to-body ratio in the animal world. While some populations have been observed to use stones to break hard nuts in the wild, it appears that tool using is not a part of their general behavior. However, in the laboratory the cebus has impressed

experimenters with a tool-using proficiency that surpasses all but the great apes.

In 1993, an Italian team led by Maria Visalberghi did an exhaustive study of cebus tool-using capacities (Visalberghi 1993), based on a battery of different tests taped by video cameras. This enabled them to do a detailed, frame-by-frame analysis of the behaviors exhibited by these animals.

The tests were generally of a similar type: a peanut would be placed in a tube, the shape of which was designed to prevent the monkey from reaching it without a tool. The experiments would vary the problem by changing the size of the openings of the tubes or the lengths and shapes of the tools available, forcing the monkey to find the right tool and use it properly in order to extract the peanut. Befitting their reputations, within a few hours, half of the cebus tested had solved all of the tests. Some were able to do so in minutes.

Careful studies of the videotapes, however, revealed a number of surprising features about the process by which the cebus arrived at correct tool-using solutions. In approaching a new problem, the monkeys would attempt many inappropriate behaviors. They would put everything they could find into the opening: pieces of tape, splinters, or whatever was lying around in the cage. Even after they had found a solution, they would still periodically attempt totally inexplicable actions. When faced with experiments that were similar

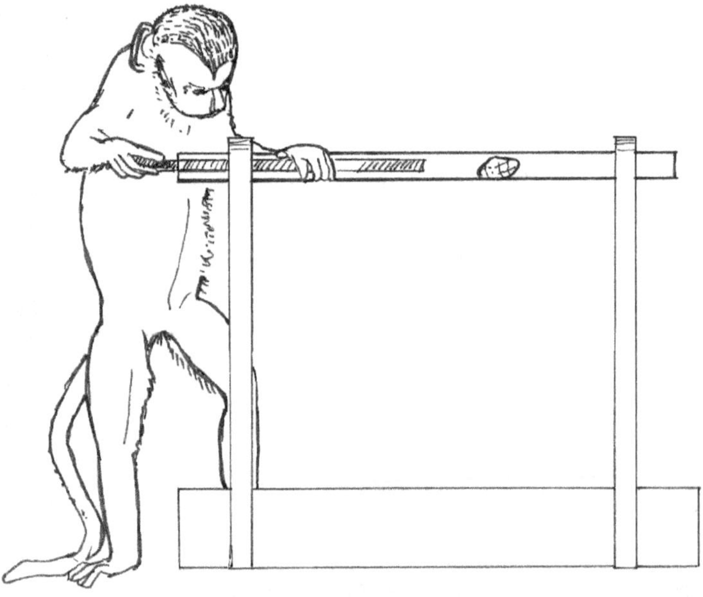

yet slightly different, they were unable to grasp the principle of the earlier solution and apply it to the new situation. Another interesting feature is that these monkeys were unable to learn from one another. When placed in a cage with another monkey that knew how to solve the problem, the monkey that was unfamiliar with the solution was unable to acquire it by observation.

In summarizing the conclusions of the three-year study, Visalberghi states: "Capuchins were successful tool users not because they understood exactly what the task was about, but because they tried every possible way to solve it, using both their own bodies and any external objects they came across" (ibid., p. 140). The cebus was unable to understand the principle involved in solving the problems, and "did not know why a successful technique was more appropriate than unsuccessful attempts" (op. cit.).

This failure to grasp the principle, she believes, is at the root of the cebus' inability to imitate the solutions achieved by others. Because it was unable to understand the problem, it could not "appropriately select what to look at in order to learn a new behavior" (ibid., p. 142). She concludes: "capuchins achieve success by trial and error attempts which are not guided by knowledge of the conditions that are necessary to achieve solution" (ibid., p. 145).

The Chimpanzee

The chimpanzee is another story entirely. While there have been many excellent recent studies of chimp tool use, the most informative, in terms of the issues we are looking at, were performed by Wolfgang Kohler at the start of the nineteenth century. Kohler, one of the founders of Gestalt psychology, carried out his research at the Tenerife Anthropoid Station from 1914 to 1917. A German national, he was cut off from his homeland by the British navy after the outbreak of WW I. He put his enforced isolation to good purpose.

Kohler had gone to Tenerife to test the mental abilities of non-human primates. He wanted to know whether they were able to solve problems through the use of "insight" (the German term he used is equivalent in English to either "insight" or "intelligence"). His strategy was to place animals in situations where they had to understand the problem in order to solve it. To do this, he designed his experiments to incorporate a level of complexity that would eliminate the possibility of trial-and-error solutions: "the more complex the problem to be solved, the less the likelihood is there that it will be solved solely by chance" (Kohler 1976, p. 16). While he began by testing a number of different kinds of primates, he soon recognized that the chimps had abilities that were lacking in the monkeys.

Kohler confronted his chimps with a number of challenging tests. In one particularly illuminating experiment, he placed a large box inside an even larger room. The box was boarded up on three sides, while the front had bars placed at intervals that would allow the chimp's arms, but not its body, to get through.

A small hole was made in the back of the box: large enough for the chimp put its arm through, but positioned too high for it to reach the floor of the box. A banana was placed in the rear of the box. It could not be reached either through the bars in the front or from the hole in the rear. The only other object in the room was a pole that had been placed against the wall behind the box. To make matters more difficult, Kohler tied a rope to the pole, which he then anchored to the wall.

Kohler had thought this experiment out very thoroughly. To reach the banana, the chimp would need a tool. Kohler put the tool (pole) behind the box so the chimp could not see both the goal (banana) and the tool at the same time. To find the tool, the chimp would have to take its attention off the goal. Kohler wanted to see if the chimp could keep the goal in mind while it searched for a means to acquire it.

Finding the pole, however, would not be enough to enable the chimp to solve the problem. Normally, the chimp would grab the pole, run around to the front of the box and use the pole to pull the banana out. Because the pole was secured to the wall, the only way the chimp could use the pole to get the banana would be to put it through the hole in the back of the cage and then push the banana away from itself toward the bars at the front. The animal would have to distance the banana from itself before it could attain it. Kohler reasoned that an animal would not do that unless it understood how the parts of the puzzle fit together. It would have to work out a strategy—solve the problem in its head before it would be able to solve it in practice.

The chimp approaches the problem very differently than

the cebus. It sits at the front of the cage and assesses the situation. It understands that the banana lies beyond its reach, and makes no effort to get it by reaching through the bars. This is typical, according to Kohler. He notes that chimps "very seldom are seen to attempt anything which would be considered accidental in relation to the situation" (ibid., p. 191).

At a certain point, the chimp looks away from the banana, and begins to survey the box and the room. Eventually it spots the pole lying against the back wall. It immediately picks it up and attempts to carry it to the front of the box, at which point, it discovers that it is attached to the wall by a rope. It looks at the rear of the box, sees the hole, and puts the pole through the hole. At first it tries to pull the banana towards the hole. It soon discovers its arms cannot reach the banana through the hole. Realizing that this strategy is not going to work, the chimp stops and reassesses the situation. Then, "the whole direction of the movement suddenly changes" (ibid., p. 254). The chimp now places the pole behind the banana and pushes it away from itself toward the bars at the front of the box. At a certain point, it drops the pole, runs around to the front, reaches through the bars and grabs the banana.

On the basis of this and a battery of other tests, Kohler concluded that the chimps were distinguished from all the other animals he tested by an ability to work a problem out in their heads without having to rely on trial and error methods. Kohler did not speculate about what goes on inside a chimp's head—that is, what distinguishes chimpanzee mental processes from those of creatures that rely on trial and error. But any observer, he tells us, can spot the difference: "a successful chance solution (trial and error) consists of an agglomeration of separate movements, which start, finish, start again, remain independent of one another in direction and speed" (ibid., p. 265). An intelligent solution, what Kohler called a "genuine achievement," however, "takes place as a simple continuous occurrence, a unity, as it were, in space as well as time" (op. cit.). Kohler went on to conclude: "Chimpanzees manifest intelligent behavior of the genuine kind familiar in human beings" (ibid., p. 265).

Kohler looked at the special tool-using abilities of the chimps in terms of a capacity to work problems out in their heads—problems that are too complex to solve through trial and error methods. Basically, he was right but this way of looking at the issue needs a qualification.

A reliance on trial and error does not mean that an animal approaches problems blindly—without intelligence. Every animal is able to mentally anticipate the results of certain actions and events. The cockroach can predict the trajectory of your hand and direct itself toward the nearest safe exit. When a raccoon attempts to break into a chicken coop, it doesn't thrash around blindly, randomly testing its various powers. It looks for doors, latches and structures that provide likely entry points.

The difference between chimps and other animals is not that chimps alone can solve problems in their heads. Rather, it's the kind of problems they can solve that distinguish them. Chimps are able to process a level of complexity that almost no other animal can.

The cognitive anthropologist Kathleen Rita Gibson (1990) has proposed a valuable way of looking at this. She argues that all the higher vertebrates (birds and mammals) are able to form highly detailed and accurate mental images of the different objects they perceive in the world around them. However, all but a very few can only see these objects in relationship to themselves: what we might call "subject-object" relationships. They are able to predict what will happen if they act on those objects and, conversely, if those objects act upon them. The cat knows it can open the door by pushing against it. It also knows to get out of the way when a door is slamming shut.

However, only a handful of species are able to predict what will happen as a result of one object acting on another. The cat cannot foresee how pushing against the chair will knock the lamp over. This involves what Gibson calls "object-object" relations. The vast majority of animals simply cannot process that level of complexity; they cannot grasp how things outside themselves are connected to other things. Only animals with the very largest brains, Gibson notes, are able to do that.

Many animals can make an "association" after the fact. The

cebus was able to learn from experience how one object could affect another, that is, by observing the results of trial and error. It was able to recognize that a certain shaped tool could move a peanut when inserted into a certain shaped tube—even if it couldn't fully grasp the mechanics of the process.

Gibson's model gives us an important insight into the cognitive threshold reached by the great apes. An ability to see how objects are related to one another and not simply how they are related to the actor is the fundamental prerequisite for understanding cause and effect. A true tool-using ability requires a capacity to understand how different objects can affect one another.

Reasoning

Kohler's experiments demonstrated that chimps could figure out how to solve certain problems that were too complex to solve with trial and error methods. He believed that on some fundamental level they went about it in the same way we do. Chimps do not begin, like the other animals he tested, by immediately trying out different, disconnected stratagems. Kohler's chimps began by studying the problem and were, as far as he could determine, carrying on some kind of internal mental process. They appeared to be "thinking" about the problem. Then, at some point, they would execute an action plan. If that plan didn't work, they would stop and "think" about it again, and then try something different.

The word that comes to mind to describe what the chimps were doing is "reasoning." That is the term we use to describe the way we solve problems through an internal mental process. Kohler did not speculate about what was going on inside the minds of his chimps; he made no effort to describe the mental processes they used to solve problems. He assumed though, that the cognitive capacities they demonstrated shared something importantly in common with human beings.

A Critical Threshold

Great ape intelligence plays an important part in the evolutionary hypothesis presented in this book. Human evolution represented a radical departure from the evolutionary dynamics that operate elsewhere in the animal world. That raises a problem. How did our early ancestors, who were the products of a conventional evolutionary process, get on a radically different kind of evolutionary track?

Evolution, at least so far as vertebrates are concerned, is fundamentally a continuous process. It's not the result of dramatic genetic changes in individual animals. It takes place through the transformation of communities by means of differential reproduction. No change in an individual can be so drastic that it is incompatible with the reproductive system of other members of the community. Populations, then, evolve together. This fact gives the process a necessary incremental dimension. The problem we face with human evolution is understanding how a continuous, incremental process could have given rise to a radical discontinuity.

There had to have been a bridge, something that could connect the two very different evolutionary trajectories. Great ape intelligence is that bridge. Human evolution took shape as our ancestors began to adapt to their surroundings with constructs they created rather than by changes in their biology. The fundamental prerequisite for this kind of behavior is an intelligence that can understand cause and effect and grasp the principle of problems that are too complex to be solved by trial and error. This intelligence, then, was a prerequisite for this behavior. It would have had to develop first—as a necessary pre-condition for human evolution.

This is not the conventional evolutionary narrative. Generally, human evolution is believed to be the product of some dramatic cognitive rupture with the animal world. The argument here, however, is for intellectual continuity. The intelligence that made the human evolutionary path possible had to have developed first in the animal world. That intelligence is what we saw demonstrated by Kohler's chimps.

However, as smart as the great apes are, it is important to bear in mind that only one of their branches was able to make cultural behaviors the basis of its survival strategy. Chimps can use tools—but not well enough to replace their reliance on their own biologically evolved resources. Tool use is not critical enough to their survival that natural selection would have targeted it. There is no evidence that the chimp organism has been shaped in any way to enable them to make better use of tools. This is where we see the important role played by the australopiths. They had occupied the margins of the savanna for at least a million years and had evolved a bipedalism that freed their hands from locomotion. The evidence suggests they had been habitual tool users throughout the better part of their species' existence. Even with all that going for them, it still took an ecological crisis to push them over the edge into a dependence on this new, more complex behavior.

Intelligence

Our investigation has led us into the minefield of intelligence. It's a place no one wants to be—because no one really knows what it is. There is no scientific consensus about how to even define the term, much less an understanding of how the mind performs its miraculous feats. Nevertheless, there is no dodging this issue. We have to address it. Intelligence lies at the core of what distinguishes human beings from all other life forms.

There are two moments in human evolution when an understanding of the nature of intelligence is absolutely critical for making sense of our evolutionary process. The first was at the beginning, when the australopiths were able to chart a new evolutionary course. The second was at the end, when changes in the human brain made possible an explosion of creativity and enabled the emergence of a higher intelligence. This was when humanity truly stepped out from within the circle of the animal world. Without a framework that can enable us to understand the cognitive events that took place during those two critical periods, it is not possible to truly understand what kind of being we are.

I wrote my PhD thesis on the role of culture in human evolution. During that process, I was able to develop many of the ideas I present in this work. However, I eventually ran into a stone wall. I was not able to develop the analytical tools that could allow me to understand the cognitive issues that were at play during those two critical periods. Without that, I could not really define what was distinctive about human intelligence. At that point, I realized that an intelligent being could not really know what it is if it doesn't know what intelligence is.

In the next three chapters, I am going to use a model of intelligence I have worked on for decades to sort out the cognitive issues involved in the first critical period—how great ape intelligence was able to bridge the gap between the human and animal worlds. A bridge is something that connects two different points; it has footings on both sides of a divide. That's what great ape intelligence is. On the one hand, Kohler was right: great apes function at a cognitive level that shares something important in common with us. At the same time, it is different from ours and gives us insight into a lower form of intelligence: that from which great ape cognition developed. To truly understand what intelligence is, we need to see it from all these perspectives—how it operates in lower animals, great apes, and human beings.

The most problematic aspect of this model is that it is based primarily on a framework extracted from human cognitive processes. That, however, is not by choice. There simply is no other way to approach this problem. We don't have a direct window into animal minds. To understand their mental processes, we have to deduce them from their behavioral capacities—as Kohler did. While that can tell us what they can do, it can't tell us how their minds enable them to do it. The only real window we have to any mind is human language. Language is materialized thought; mental processes that are given a tangible form (sound, signs) so that they can be shared with others. Writing takes this a step further. It is petrified thought: thought that is given a permanent form and that can be subjected to thoroughgoing analysis. It is no accident that the ancient Greeks, who were the first people to invent a fully developed alphabetical system, were also the first to develop a theory of the mind—philosophy. Their discoveries gave birth to a philosophical tradition that dominated Western intellectual life for over two thousand years—and from which I gained a major portion of my ideas about the nature of intelligence.

Only a few years ago, it would have been unthinkable to use the human mind as a template for making sense of animal intelligence. It was generally assumed that animal and human mental processes operated by entirely different principles. Looked at from an evolutionary perspective, however, it is clear that is exceedingly unlikely. The great ape brain is the product of 450 million years of vertebrate evolution. Our brains, which are built on a great ape biological platform, are the result of only 3 million years of a specifically human evolutionary process. It would seem to be impossible that their brains would work by entirely different rules than ours.

A Brief Introduction to the Section on Philosophy

In the next chapter, I am going to draw upon Western philosophy's model of the human mind. I believe it gives the clearest picture of how we humans go about solving the kind of problems that Kohler's chimp faced. It provides us with a framework that enables us to make a connection between our own mental faculties and those of the great apes. But the chapter is a difficult hurdle for anyone who is not familiar with philosophical concepts.

Philosophy operates on a very abstract level; it's the nature of the beast. Its subject is the mind, and the fundamental principle of the mind is abstraction. The mind's job is to simplify the world—to discover patterns and commonalities that make it predictable and that enable living creatures to develop behavioral strategies. It accomplishes this by means of abstraction—by setting aside the specific features that distinguish things from one another and focusing instead on what things have in common.

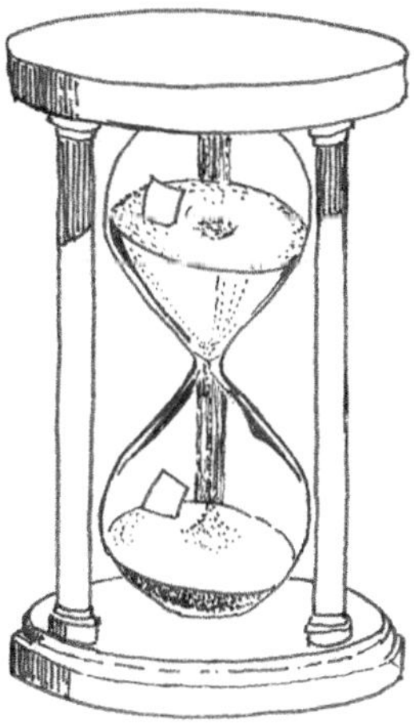

Chapter 4

Reasoning: Western Philosophy's Model

Early in my research, I had to face the fact that the contemporary cognitive sciences have not yet developed a comprehensive model of how the human mind works—much less a more general model that would allow a comparison between human and animal intelligence. There are a lot of good ideas, and the neurological sciences are making great progress, but nothing exists that could provide answers to the questions I was asking. In desperation, I turned to what might seem a most unlikely source—Western philosophy.

That decision, however, was not entirely capricious. The great apes, as we saw, demonstrate a capacity to work certain kinds of complex problems out in their heads. When this capacity is exhibited by humans, it is called reasoning. Western philosophy has been investigating the nature of reasoning for 2500 years—and some of humanity's greatest minds put their all into trying to fathom this issue. It would seem that there should be something of value there. And there is.

I decided to begin my research by looking into early modern Western philosophy—the works of Hume, Kant and Hegel in particular. I soon discovered, to my great good fortune, that I had stumbled upon a debate that addressed the very issues we examined in the last chapter: cause and effect, trial and error, and reasoning.

Cause and Effect

The discussion began with the Scottish philosopher, David Hume (1711-1776). Hume was a leading thinker in a movement that was challenging certain traditional dogmas in classical Western philosophy—ideas that were increasingly at odds with a newly emerging scientific world outlook.

Classical philosophers, going all the way back to Plato, had disparaged the senses as limited and unreliable sources of information. They held that the higher faculties of the mind—those that could grasp the underlying logic of events—were the only true path to knowledge. This led to a situation where ungrounded and untested philosophical speculations, as long as they were internally consistent (logical), could be accepted as valid knowledge.

Hume turned the argument around. He proposed that the information we receive from the senses—not logic—is the real foundations of all knowledge. The senses are our only direct connection to the external world, and they alone provide us with information that we can be certain of—that can be confirmed. And the senses, he insisted, cannot tell us how one thing can be the cause of another.

The senses, Hume argued, can only tell us that two objects or events are associated in time and space. They can tell us that one event came before another and that two things were in contact with each other: they cannot tell us that one was the cause of the other.

We can see that the man ate the mushrooms and then became sick and died. That information was given to us by the senses. They inform us that one event—eating the mushrooms—was followed by another—the death of the man. They can only tell us that these two events were associated with one another; they cannot tell us that the mushrooms killed the man. We can come up with a theory about how the mushrooms might have killed the man, but we do not have direct sensory evidence that can verify that. We could argue, of course, that chemists have isolated a toxin in the mushrooms and that this caused the man's liver to fail. But Hume is always able to stay one step ahead of this line of argumentation. All that the chemists can know is that when this toxin is administered, it has been associated with the destruction of livers. We cannot know for certain—the senses cannot confirm—that this toxin was the cause of death.

Hume is generally credited with injecting a dose of healthy skepticism into a human tendency to jump to unwarranted conclusions. But in fact, his argument raises an issue that has profound implications for the evolutionary issues we are discussing.

Let us go back to the experiment with the *Cebus* monkey. The *Cebus* possesses senses that are more less equal to those of a chimp, yet it could not "see" the problem the same way the chimp could. It was unable to recognize how the shape of the tool and the shape of the tube had to match up in order for the tool to be able to push the peanut out. Unable to process such complexity, it was forced to resort to trial and error.

When its trial and error methods were successful, when they revealed a connection between a particular tool and a particular tube, the cebus could recognize the connection—even if it could not accurately model how the effects were produced. This is association. What is it that the *Cebus* doesn't get? It's senses—eye sight—are basically the same as those of the chimp. But it doesn't see the problem in the same way.

Before this work, I had never understood the acclaim that Hume has received as a philosopher. His contention that we can never really know what in fact we are able to know—how a chain of events produces certain results—always seemed to me more like a contrarian parlor game than a serious intellectual challenge. However, in trying to make sense of the difference between the problem-solving abilities of the cebus and the chimp, I have become convinced that Hume was raising a real issue—and that on one level he was right. The senses cannot grasp cause and effect. An intelligence that depends primarily on sensory information can only perceive an association between events; it cannot provide a creature with a model of a causal process. That sounds bizarre. Not only humans, but great apes—as well as a few other animals—are able to understand causal processes. But there are two different issues here. What Hume is saying is that the senses can only perceive association, not causation. I think he is right about that. Where his argument goes astray is in claiming that all we can know is what the senses tell us.

The human mind—and those of the great apes and some other animals as well—is able to go beyond the information provided by the senses. Hume denied this dimension, and in so doing, substituted one dogma for another. It would be more than 40 years after Hume published his provocative thesis before someone would be able to develop a thorough-going critique of his ideas.

Immanuel Kant

Hume's challenge was taken up by one of philosophy's greatest thinkers, Immanuel Kant (1724-1804). In his *Critique of Pure Reason*, Kant subjected Western philosophy's theory of knowledge to a deep-going reassessment—and in the process, put together an original model of how the mind works.

Kant began by agreeing with Hume that the senses are indeed the only direct access we have to the external world. However, Kant goes on, they are not the only source of information we have about it. How is that possible? How can we know something about the world that does not come by way of direct experience with it?

The mind, Kant argues, knows something about the world that the senses don't. It knows that the world is logical. And it knows this previous to all experience—what he called a priori. It is an assumption that is built into the mind.

Model Building

What Kant is arguing is that the senses do not give us the whole picture. They are an important, but also limited source of information.

While the senses can inform us that a shadow is extending over the Moon's surface, they cannot tell us what is producing this effect. To be able to understand what's causing the shadow, that it is the result of the Earth coming between the Sun and the Moon, we need more than the evidence the senses can provide. We need a model—a way of visualizing how these celestial bodies orbit each other. Such a model is not given to us directly by the senses. It is constructed out of the incomplete evidence they provide. That's where logic and the higher faculties of the mind come into play.

Copernicus' senses could not inform him the Earth revolved around the Sun. Indeed, they suggested the opposite. He was able to construct a heliocentric model of the solar system because it provided a more efficient way to organize all the bits and pieces of evidence that the astronomy of his day had collected.

There are two parts to this. The first is that our models of the world are not given to us directly by the senses. They have to be constructed. And secondly, that in building those models, the mind is guided by an '"instinct" that the different pieces of information we have should fit together to form an internally consistent, coherent whole. That's logic.

Kant's view represented a revolutionary new outlook for philosophy. The idea that our models of the world are constructs is not only different from Hume, but it's also different from the way the classical philosophers understood knowledge. They believed that logic was the organizing principle of the universe and that the task of the mind was simply to uncover or recognize that design. Kant was saying something different. He proposed that the reason we discover logic everywhere we look is because our minds are built to process information into logical models. Logic, then, is a property of the human mind, not necessarily of the world.

Kant agreed with Hume that the senses cannot provide a creature with a model of cause and effect. The human mind, however, is not limited to the information supplied by the senses. It processes that information and uses logic to fill in the gaps left by sensory experience. Our models of things—the way we understand the world—are inferences. While they rest on sensory information, they go beyond it. It is this ability to go beyond the evidence provided by the senses that enables us to work out complex problems without resorting to trial and error.

What does all of this have to do with chimpanzee tool use? A lot. In the experiments we looked at, the *Cebus* could only recognize association: that certain actions were accompanied by certain results. It couldn't form a model of how those actions produced those results. The chimp could do that. It could model the process and thus grasp the "principle"—while the *Cebus* couldn't.

Chimps, and all the great apes, have passed over a cognitive threshold. They are able to form models of causal processes and they can solve certain complex problems without having to resort to trial and error. The *Cebus*, I believe, represents an intelligence that is based primarily on sensory information. The great apes are able to process that information and take it a step further. This is the threshold that provided our australopith ancestors with the intelligence to begin performing more complex behaviors. And that intelligence, I believe, shares something fundamental with our own.

At this point, I would like to take a deeper look at the concept of logic.

The meaning Kant (and Hegel) gives to this term holds the key to understanding his philosophical system.

Logic and Systems

We tend to think of logic as a set of rules—guidelines—for organizing one's thinking. That's what the term has largely come to mean today.

Hegel and Kant saw it as something much more than that. They saw logic as a fundamental assumption that the mind makes about everything. It is a template the mind uses to construct all its models. (See appendix on the difference between Hegel and Kant.)

What is this assumption? That everything is a system. The way they put it was that everything is simultaneously a "One and a Many." That's an arcane way of describing a system. Think of an ecological system. It's a whole—a One—that is made up of a vast network interdependent parts—a Many. Change one part—remove the wolves—and all the other parts of the system are affected.

The philosophers believed the mind is designed to look at everything in these terms—as a system. Indeed, unless information is organized in this way, the mind will not find it "intelligible." To understand something means to see how the parts of a problem, object, or event fit together to form a single, coherent whole.

Think about it. Everything you "know," everything you understand, is something you can describe as a system—a whole that is made up of interconnected parts. This is true of every object or entity. A house is a whole made up of a number of different parts: a roof, sidings, plumbing, rooms, foundation, etc. The same holds for a wheelbarrow, a desk, a stove, a computer, a tree, a flower, an internal-combustion engine, a government, etc. You cannot bring to mind any object that you do not model in these terms.

This way of organizing information applies not just to objects but also to events. Indeed, while we don't normally recognize it, the human mind looks at objects and events in basically the same way. While an object is a system in space, an event is a system in time—a process. Every event you can think of—your daughter's wedding, a trip to the store, last year's election, World War II, cleaning the drainpipe—is a whole made up of different interconnected parts. Only here the parts are sequential. An event is a whole—a One: an activity that is bounded by a beginning and an end. At the same time, it is a Many; it is made up of a series of interconnected steps or episodes (parts).

We don't think of this way of looking at things as something we do: a product of our mental processes. We think this is just the way the world is. It is only when we see that other creatures can't see things this way that we become aware that our way of modeling the world is a construct—the product of a particular level of intelligence. It is the way we process experience.

Cause and Effect and Reasoning

If we can represent something as an interdependent system, then we can see how the different parts of that thing are related to one another. In the previous chapter, I discussed object-object relations and how few animals can fully grasp how things in their environment are connected to one another. That is because they cannot fully process their environment as a system. The cat could not foresee that pushing against the chair would knock the lamp over because it could not process how the different pieces of furniture were connected to one another. As a result, it could not predict how pushing against the chair would affect the lamp.

Being able to model things as systems (or processes), that is, to see them as a set of interconnected parts, is the basis for understanding cause and effect. There is a video of an orangutan studying a door latch—examining how the different parts are connected to each other. It is building a model of the latch: figuring out how it works. This is what Hume didn't get. We—and the great apes—can construct models that allow us to visualize causal processes. The orang did not experience pressing its thumb on the lever and the withdrawal of the deadlatch as two disconnected events. It experienced them as two inseparable parts of a single interconnected process—opening the latch. And as the caretakers on the orangutan reserve learned, it's very hard to design a latch that can keep an animal that can think that way out of the pantry. The ability to represent things as systems and grasp cause and effect relations are inseparably connected to the capacity to solve a problem in one's head—what I have called "reasoning." If we can see something as a system or process, how the different parts of a problem are related to one another, it becomes possible to predict how affecting one part of the system will have an impact on other parts.

When we reason, we are system building: attempting to put different pieces of some problem together into an organized, coherent model. That's what Copernicus did. He showed that placing the Sun, rather than Earth in the center of the solar system would resolve certain anomalies in the movements of the planets. They could now be seen to form a coherent system of orbits. It's how Sherlock Holmes solved his mysteries. He assembled the different clues and looked for a thread that could tie them together—and point to the culprit.

When we deduce, that is, seek to discover an unknown, we do it by completing a system. If we can look at something as an interdependent system, then we can figure out what the missing piece or pieces most likely consist of. By looking at an internal combustion engine as a system, we can diagnose a problem through a process of elimination. If the gas tank is full, we have plenty of oil, and the spark plugs are new, the problem almost certainly is the carburetor. By recognizing that the planet Neptune's orbit was not in line with the orbits of the other planets, astronomers were able to deduce that another, but as yet unperceived body, must lie beyond it.

This is the way we solve problems; the way we use what we know to discover what we don't.

Logic: The Instinct of the Mind

The dominant trend in the cognitive sciences today is to see the mind as a kind of computer. Data is taken in and processed into models that enable us to function on a human level. Superficially, this sounds a lot like what I am suggesting. But in fact, it's very different. The computer model sees all this happening automatically: outside the framework of conscious deliberation. And, of course, there is an important automatic dimension to the workings of the mind. But that's not what reasoning is. The chimp's mind did not automatically generate a plan to retrieve the banana. The chimp had to "think" about the problem; it had to look at the situation and put the different pieces together in its mind. It had to create a plan—a conscious model of the how it could get the banana. And when the first plan didn't work, it had to re-evaluate and generate another.

However, while the reasoning process is not automatic, it is guided by a built-in compass. That's the way the philosophers saw logic. We have to put the pieces of the problem together, but the process is directed by the expectation that they must ultimately fit together. The mind will not accept an explanation or model in which the parts are not consistent with one another and don't form an integral whole.

When we "connect the dots," that is, put the different elements together to form a coherent narrative, we get an "aha" moment: a sense that we understand the problem. Understanding is the mind's way of telling us that we have organized the information into a system: that we have given it a logical form.

However, the fact that something is logical, that the parts fit together to form a coherent story, does not mean it is true; think about all the wacky conspiracy theories that are currently circulating. It does mean, however, that it meets certain prerequisites of plausibility—logical consistency.

This discussion has led us to the borders of philosophy's most profound problem: Is the world logical—or just the mind? That's the place where Hegel made his mark; but that discussion is for a different book (see appendix).

Conclusion

There are two important conclusions that flow from the philosophical framework we have examined. The first is that "reasoning"—the ability of humans to understand cause and effect and to work complex problems out in their heads—is based on an ability to model objects and processes as systems. The second is that the great apes have reached a mental level where they can begin to do this. Their abilities in this regard, however, are very primitive in comparison to our own. Nevertheless, they are a step above almost every other animal. This was the level of intelligence that enabled the australopiths to initiate more complex kinds of behavior.

Unfinished Business

In this chapter I have used the framework developed by Western philosophy to understand the tool-using capacities demonstrated by great apes. The fact that human cognitive processes could shed light on this problem suggests that there is a common denominator between great ape and human intelligence. This should not come as a surprise—indeed, it is to be expected. We evolved from a great ape and that great ape had to have been able to perform certain kinds of complex behaviors in order to have put our lineage on a new evolutionary track.

This analysis, however, leaves us with a number of problems. The first is that it rests on shaky ground. To use a framework that has been inherited from humanity's pre-scientific past to establish a connection between great ape and human intelligence is quite a stretch. I believe Western philosophy is a lost treasure and that, at this point in our knowledge, it provides the best foundation for understanding the questions I am asking. But I also recognize we need more and different kinds of evidence. It is not possible to develop a fully-rounded understanding of the issues we are looking at without drawing on some of the latest discoveries in the neurosciences.

There is also another problem. This is a book about human evolution—about placing humanity into the context of the natural world. While it is critical to recognize a connection between human and great ape intelligence, it is also important to understand the connection between great ape intelligence and that of the animal world out of which it evolved.

In this chapter we looked at great ape intelligence from the perspective of human intelligence; what it has in common with our own. Now I want to turn the lens around and look at great ape intelligence within the framework of the animal world. Great ape intelligence is transitional. The apes have one foot—or at least a toe—in a higher intelligence. But they are barely there; they lack language and in most ways are closer to the mental level of other higher vertebrates than they are to us. The transitional nature of their intelligence indicates a process—something becoming something else. The higher intelligence that is emerging is growing out of a lower one.

In order to understand the relationship of our intelligence

to that of the animal world, we need to understand the cognitive developments that took place in the great apes. Those developments represent the first steps in a process that eventually led to us. To understand that process, we need to know where it began and what happened with the great apes—what changes in their brains elevated their intelligence and opened up a new evolutionary pathway.

Chapter 5

The Vertebrate Brain and the Emergence of the Self

The idea that there is something different about the mental abilities of great apes and humans immediately raises the question—different from what?

There is a tendency among many scholars to treat a reasoning capacity as if it were synonymous with intelligence. This assumption relegates the behavior of other animals to something called "instinct"—built-in action responses that involve no intellectual process. Instinct is a useful term, and it can be applied to certain aspects of animal behavior; indeed, even human behavior is importantly shaped by responses that are built into our organisms. But living creatures are not automatons. Intelligence enters into the behavior of every animal.

I use the term intelligence very broadly—to describe the capacity of a being to gain an awareness of its world and to use that awareness to interact with the objects and other beings that inhabit that world. Animals do not stumble around blindly. They act with intentionality; they have purposes and they are able to make choices about how to realize those purposes. And most importantly, they can learn. Learning requires intelligence—an ability to connect acts and events with their consequences. While the associations made by the cebus may fall short of grasping "the principle," nevertheless, they require an act of intelligence. Even lowly earthworms are able to learn the right way to turn in a T-maze in order to avoid an electric shock. This very broad definition of intelligence, then, includes everything from amoebas to us.

The terms reasoning and intelligence, however, are not synonymous. Reasoning is a specific form of intelligence: a "higher" form, if you will. This distinction raises the question—what does intelligence without reasoning look like? To be able to understand what reasoning is, it is necessary to also understand what it is not. How could we describe an intelligence that cannot reason—the intelligence out of which the mental capacities of the great apes evolved?

The Structure of the Vertebrate Brain

In an article entitled "Origins of the Left and Right Brain" (*Scientific American* July 2009), a team of scientists announced the results of some exciting new research in comparative neuroanatomy. The research established that the brains of every kind of vertebrate (i.e., fishes, amphibians, reptiles, birds, and mammals) have a common design feature. They are all constructed in a way that provides a creature with two different kinds of information about the world. The article gave a brief summary of an extensive body of new experimental data that appears to corroborate the old split-brain theory that achieved a certain popular cachet during the 1960s and 70s.

Animals, the authors propose, need two different kinds of information in order to function. On the one hand, they require specific information about the different objects and events they have to deal with. They need to know what "type" of tree, nut, soil, predator, liquid, etc., they are facing. They also need to be able to discern the specifics of a particular individual object—its size, color, odor, state of mind, etc.

Besides specific, concrete knowledge of particular objects, a creature has to be able to grasp relationships—both in time and space. It has to know the "lay of the land," that is, be able to place objects into a spatial context in order to understand how things are positioned in relationship one another. It needs to recall where sources of food and water are located. And most importantly, it has to have to have a mental map of various escape routes.

The experiments confirmed what had already been widely acknowledged. The left hemisphere of the brain focuses on objects, organizing information about them and providing the animal with certain routines for coping with them. The right hemisphere, on the other hand, does not focus on specific objects, but rather, supplies a creature with a global sense of a scene or event. It "intuits" a field: it sets the general boundaries of an object or event, but doesn't provide details of what is connected to what or how they are connected.

One side of the brain deals with specific things— focuses on particular objects (the Many). The other side deals with global patterns—the larger context—providing a frame for seeing how the individual things are related to one another (the One). The merger of the two different sides creates an integral field of experience. What emerges from this analysis is a model of a mind that can focus on specific things while maintaining a sense of the larger context in which they operate (their place in a system).

This new neurological research would seem to lend weight to Western philosophy's accomplishment—providing evidence of a connection between the model of the human mind it developed and the organization of the vertebrate brain. What the philosophers called the "instinct of the mind"—logic—is an expression of the basic organizational principle of the vertebrate brain: the rules the mind uses to construct a system.

This analysis, however, leaves us with the problem I want to look at in this chapter. What's different about a mind that can reason? If the brains of all the vertebrates have the same basic structure, and all of them are designed to organize experience into systematic models, then why can't every vertebrate reason? Why can't all of them process object-object relations and work out complex problems in their heads? What do those who can do this have that those who can't lack?

Size Matters

The intelligence of an animal is not determined by the general design of its brain alone. One of the most distinctive feature of the great ape brain is its immense size. The brain of a chimpanzee, in relation to the weight of its body, is three times that of the average mammal (cats, dogs, horses, lions, etc.). Its brain is thirty times the relative size of that of a fish, amphibian or reptile. The primate order in general has the largest brains in the animal world, and the great apes have the largest brains of all the primates —except for us. The size of the brain is clearly an important factor in intelligence. It provides more space for memory and is able to house a more extensive neural network for processing data. It should come as no surprise that animals with larger brains perform better in intelligence tests.

Gibson was right: all animals that can process object-object relations have very large brains. The other animals that demonstrate this capacity—the corvids (ravens, crows, New Caledonian crows, etc.), some parrots, porpoises and some other toothed whales, and perhaps a few other animals as well—all have exceptionally large brains. In many cases, they fall within the range of the great apes.

But it's not just about size. The brains all of these intelligent animals possess have another distinctive feature: a large cerebral cortex (in birds, it's a different body, but it performs the same function). The cortex is the most recent part of the brain to have evolved, and the lower vertebrates—fish, amphibians and reptiles—do not have a cortex. It does not simply receive and process sensory information, but rather, plays an integrative role. It connects the different sensory regions together to produce richer and more complex perceptual models.

This difference in size and composition would seem to give substance to the debate between Hume and Kant. A brain that either lacks or possesses a small cortex, generates models that are built almost entirely out of primary sensory inputs (*Cebus*). A brain with a large cortex is able to add something to those inputs; it is able to use logic to deduce a more complex model of its environment (chimp). I think there is something to this; but the picture is still more complicated than this.

Agency

Kohler was not only impressed with what the chimps could do, but also how they went about it. In order to solve the tool-using problems, they had to go through an internal mental process—what I have called reasoning. Reasoning is not something that happens automatically—something the brain does for us. It is something we do. When we reason, we are seeking to improve on the information that has been given to us by the senses. We are looking for connections that are not given directly in perception. We see a big boulder in the middle of the road. Our senses can only tell us that it is there—not how it got there. For that we need additional information. We look at the hillside next to the road. Are there divots and displaced earth that might mark the path made by a rock rolling down the hill? If not, we look for skid marks on the pavement; perhaps it fell off a truck. When we reason, then, we are trying out different frames (hypotheses) in an attempt to discover a narrative (process) that might account for the facts or events we seek to understand.

Reasoning, then, is a conscious, deliberative process. But that raises the question; Who is deliberating with whom? What enables a mind to critically evaluate its own models—to be able to recognize the need to go beyond the evidence provided by its perceptual processes?

One of the more thoughtful approaches to this issue is that of Antonio Damasio, professor of neuroscience at the University of Southern California (2010). Damasio argues that the key to a higher intelligence lies in the emergence of a second center of consciousness. This second center enables an animal to be aware not only of the world around it, but also, of its own mental processes. A creature that is aware of its own thoughts has distance from those thoughts; it can interact with the mind's spontaneous products and construct more comprehensive models.

This awareness of one's own mental processes gives rise to a self—a more autonomous being; one that is not the captive of automatic mental processes, but rather, is able to play an active part in the construction of its own mental world. The self, then, is not simply the product of a higher intelligence; it plays an active role in creating a higher intelligence. The emergence of a self was a turning point in the evolution of intelligence.

A self-reflective being has one foot outside the perceptual framework provided spontaneously by the mind. This distance is what enables it to bring other perspectives to bear on its situation—to try out different narratives (hypotheses) and be able to evaluate and modify its models.

Self-awareness, it is important to recognize, is not something that is different from the reasoning process I described in the last chapter. It is, rather, an essential dimension of it.

Intelligence without Reasoning

An animal without a self-reflective capacity possesses intelligence, but it cannot re-work and improve upon its models through an internal process. Its awareness is predominantly determined by automatic perceptual mechanisms and can only be modified through direct sensory experience—trial and error. It lives in a Hume-ian world; events are connected by associations rather than by causal necessity. The emergence of self-consciousness—of a self-reflective mind—-marked a critical threshold in the evolution of life on our planet.

This analysis of a self-reflective mind clearly corresponds with our own experience and is entirely consistent with the philosophical conclusions we examined earlier. But is this framework really applicable to the great apes? Is there any evidence that they possess a self-awareness?

Chapter 6

Self-awareness and the Great Apes

The idea that great apes might have a sense of self-awareness dates from a number of experiments with mirrors carried out by Gordon Gallup Jr. in 1970. Gallup performed these tests on chimpanzees and three different species of macaques. When mirrors were placed in front of the animals, all of them initially responded as if confronted by another animal. After a while, the chimps were able to recognize that the images were reflections of themselves. The monkeys were never able to make that connection.

Since that time, a number of different animals have been tested by a variety of researchers, and the results have produced conflicting claims. Whatever specific differences may exist, however, there is general agreement that the overwhelming majority of animals are unable to recognize themselves in a mirror. A few species beside great apes—dolphins, elephants, and Eurasian magpies (corvids) in particular—appear to be able to do this. However, most of the experiments on these animals have not been duplicated nor carefully reviewed. Nevertheless, based on the cognitive capacities they demonstrate in other circumstances, it would seem entirely plausible that they could perform this behavior.

However, many claims of mirror recognition have not withstood scrutiny. In 2003, Gallup, along with James Anderson and Daniel Shillito, revisited a number of mirror tests performed on different primates since the time of original experiments. According to the authors, re-testing showed that only the great apes could pass the mirror tests. Not even such intelligent monkeys as *Cebus* or macaques—both of which can acquire tool-using behaviors—can recognize themselves in a mirror. Indeed, while chimps, bonobos, and orangutans have little trouble with self-recognition, it is not entirely clear that gorillas can do this. The experiments also reveal that human infants cannot pass this test until they reach 18 months of age.

Self-recognition is actually an incredibly complicated cognitive feat. As the authors of the article put it: "In its most rudimentary form self-awareness is the ability to become the object of your own attention. When you see yourself in a mirror, you are literally the object of your own attention, but most organisms respond to themselves in mirrors as if confronted by another organism." (Gallup 2003, p. 9-10).

The difficulty in recognizing some object (the image in the mirror) as a reflection of oneself is that we never directly experience ourselves as objects—as something outside and different from ourselves. We are subjects; we are the ones doing the experiencing. To recognize that the object in the mirror is the subject that is looking at it, while simple for us, is a huge cognitive challenge for other creatures.

However, the idea that we never directly experience ourselves as an object needs to be modified. There is a way a being can directly experience itself as an object: when it is aware of its own thought processes. To be aware of one's thoughts is to experience oneself (the thinking self) as an object—as something that, in the words of Damasio, is being observed—something that can be distinguished from that which is doing the observing. The seventeenth century French philosopher Renee Descartes famously expressed this when he

proclaimed: "I think, therefore I am."

Besides being able to recognize themselves in mirrors, great apes demonstrate a number of other behaviors that indicate a degree of self-awareness. To begin with, they are aware of how they appear to others. This enables them to practice deceit. The ability of chimps to imagine how others see them—how they appear in the eyes of another—allows them to assume a demeanor intended to mislead an observer about their true intentions. Zoo goers are often rudely surprised when an apparently sleeping chimp jumps to its feet and hurls a handful of feces at them.

Imitation is another expression of self-awareness. While we often say "monkey see, monkey do," in fact, monkeys are unable to copy the actions of others. Only great apes can imitate. To be able to mimic another's behavior, an animal must be able to see itself as an object: to shape its own actions according to its perceptions of the movements of another.

There is one last element in great ape self-awareness we need to consider. Great apes not only have a special sense about themselves, but also about others. It appears that they alone among all non-human animals possess what researchers call a "theory of mind." This is a term that refers to the ability of a being to impute mental processes to another. A being that is able to do this can recognize not only what another creature is doing, but also, what it is thinking—the intentions that lie behind its actions. This is fundamental for complex social behavior.

The notion of a "theory of mind" has given rise to a heated debate, and, as might be imagined, has been difficult to prove one way or the other. It involves determining what an animal knows about what's going on in another animal's mind. However, a number of ingenious experiments appear to have finally settled the issue.

In an important article summarizing the evidence, Call and Tomasello (2008) give "a definite yes" to the question of whether chimpanzees have a theory of mind. When placed into situations where they had to compete with another chimp for food, "chimpanzees take into account what their competitor can and cannot see, what he can and cannot hear and even what he does and does not know." This capacity enables an individual to design its behavior on the basis of the attitudes and information it believes are possessed by another animal.

In one very clever experiment, the chimps were able to figure out what another animal (it was a human) was trying to do, that is, the goal it was seeking, even when the other animal was going about it in the wrong way. A being that is aware of its own thinking process can attribute that capacity to others: "Knowledge of self is an inductive springboard for an inferential knowledge of others" (Gallup 2003, p. 10).

Self and Other

Self-consciousness is often thought of as something inward and psychological: the ability of a being to sense its own separateness and uniqueness—to be aware of its own inner feelings and thoughts. That, of course, is part of it. However, to be aware of oneself is to think of oneself as an object— as one might appear in the eyes of another. This gives one distance from one's inner processes, allowing one to think of oneself "objectively."

To see yourself from another's perspective is to put yourself into a larger frame—one that includes the other as well. The ability of great apes to develop a theory of mind suggests that a being that can imagine how it looks to others, can also imagine that those others see things the way it does. To see oneself as an object appears to go hand-in-hand with the ability to see another as a subject—as a being that thinks like you do.

With self-consciousness, a being takes a step outside the sphere of its own organism; it moves beyond the confines of its own subjectively defined world. A self, according to Damasio, is aware that it exists within a larger world—one that includes other selves. This ability to see oneself as a part of a larger whole has vast cognitive, emotional, and social implications; it provides the cognitive foundation for cooperative behavior.

Michael Tomasello, co-director of the Max Plank Institute for Evolutionary Anthropology in Leipzig, has played a key role in focusing evolutionary anthropology's attention on the cognitive challenges posed by human cooperation. Tests show that chimp and human infants have similar tool-using capacities (object-object relations) up until around 2 ½ years of age. However, their social skills are starkly different. Even before language becomes a factor, human children are able to read each other's intentions and work toward collective goals—behavior that is far beyond the capacity of their chimp counterparts. While the great apes can perceive object-object relations, they are unable to solve subject-subject problems—to do what it takes to get a number of different actors on the same page. Tomasello concludes that these social capacities may well be due to "species-unique cognitive skills"(Tomasello 2014, p. 43)—that is, they do not have their roots in great ape intelligence.

Tomasello might be right, but I have my doubts. I don't believe it is necessary to look outside the cognitive framework we inherited from the great apes for the seeds of our social intelligence. It is more likely that our social skills evolved out of great ape self-awareness and developed as an aspect of our overall general intelligence.

In closing this section, I would like to note that Gallup and his co-authors end their article on self-awareness by providing evidence that the capacity for self-recognition appears to be centered in a particular region of the brain—the pre-frontal cortex: "There are . . . reports of brain damaged patients who fail to recognize themselves in mirrors . . . a patient with damage restricted to the right prefrontal cortex that could identify other people with a mirror, but insisted that his own reflection was not himself." The authors go on to cite other studies—all of which "implicate . . . localization of self-recognition in the prefrontal cortex" (Gallup 2003, p.12-13).

This introduces us to the next issue—the last piece of the puzzle we are trying to put together. Is there any special feature in the brain of great apes—something they share with us—that might underlie their distinctive cognitive capacities?

The Pre-Frontal Cortex

During the first decade of the twenty-first century, research in comparative neuroanatomy—comparing the brains of humans with those of other animals—really took off. One of the factors driving this research was the desire to find some feature that might distinguish the human brain: something that might account for our special intellectual capacities. It was clear that size—the extraordinary dimensions of our brains—was a critical factor. While the brain of a chimp is three times larger than that of the average mammal, our brains are more than three times greater than those of a chimp. But is the difference simply one of more neural tissue? Or are there certain unique structures that might explain some of the more remarkable human intellectual capacities? Attention became focused on a small outcropping at the very anterior part of the brain: the pre-frontal cortex. Evidence gathered from the effects of damage to this region indicate that it is "associated with cognitive capacities related to human uniqueness: purposeful actions towards higher-level goals, complex social information processing, introspection, and language" (Smaers: 2011). The expectation was that the relative size of the pre-frontal region would be much larger in humans than in other animals. That turned out to be almost correct but not quite. There was one other group of animals that also possessed an exceptionally large pre-frontal region: the great apes.

Recent research reveals that the great apes, besides having the largest brain of all non-human primates, are distinguished by the disproportionate size of the prefrontal cortex.

Definitive Study

In March of 2013, the Biological Division of the Royal Society published an extensive study that compared the prefrontal cortex of 17 different primate species (J.B. Smaers and C. Soligo 2013). According to the report, the great apes were distinguished from all the others by "the massive increases in the brain's prefrontal cortex." In apes, this region swells to four times the relative size it reaches in monkeys.

While the prefrontal cortices of apes are a good deal smaller than those of humans, that is due almost entirely to our larger overall brain mass. Relative to the size of the brain, the pre-frontal cortex of apes occupies more or less the same volume as it does in humans. "For the first time," concluded Jeroen Smaers, an evolutionary biologist at the University College London and the study's co-author, "we can really identify what is so special about great ape brain organization."

What is the Pre-Frontal Cortex?

There are two features that stand out about the pre-frontal cortex (PFC). The first is that it contains the highest concentration of white matter in the brain. White matter is composed of axons, extensions of brain cell bodies (grey matter) that serve as "wires" connecting different neurons and regions of the brain. They form the most extensive network of connections found anywhere in the brain. What could explain the larger size of this neural body in great apes and humans? The authors of the report believe it has to do with the overall size of the brain (see also: Goldberg 2009). It would seem that as the ancestors of contemporary apes began to evolve larger brains, their brains became more internally differentiated and the distance between the different centers became greater. The pre-frontal cortex most likely evolved as a mechanism to better organize and integrate the brain's more distantly separated and more highly differentiated parts. It serves as a central '"switchboard," providing an infrastructure that enables more direct and better communication between the different regions.

It appears, however, that the PFC does more than this. Lying at the heart of a vast network of neural connections, it evolved into what the neuroscientist Elkhonon Goldberg calls a "command post," bringing a higher level of organization and focus to the functioning of the mind.

The Executive Function

All complex behavior, that is, all behavior that involves linking together a number of separate steps to reach a goal, requires a plan. A plan is a mental blueprint; a consciously worked out scheme that enables a being to combine different behaviors into a single process for achieving some end

Kohler's chimp had to find a tool before it could get the banana. It had to turn away from the banana—the goal—yet keep it in mind. It didn't need just any tool, but one that could solve the particular problem it faced. It had to combine different behaviors (finding and using a tool) into one continuous problem-solving process (attaining the banana). That's what a plan does. All cultural behavior has a discontinuous dimension; it involves using something outside the subject as a means to solve a problem. Everything we do, from making a sandwich (assembling the ingredients) to building a house requires a plan. The "organ" for planning, it turns out, is the prefrontal cortex. Elkhonon Goldberg, the last of a trilogy of great Soviet psychoneurologists (Vygotsky was Luria's mentor, and Luria was Goldberg's teacher), is one of the world's most knowledgeable authorities on the nature and role of the PFC. The key to planning, he explains, is that an animal must hold two different thoughts in its mind at the same time. To be able to formulate a plan, one must keep the goal in mind while organizing the different steps to attain it.

Goldberg's description of how the PFC organizes thinking and constructs plans fits nicely with the model of reasoning presented earlier. The goal is the whole (the One). It sets the framework for organizing the parts (the Many)—the concrete steps necessary to achieve it. The plan is the synthesis—when the whole and the parts have become merged into a single scheme.

This gives us another layer in our understanding of reasoning. When we think of how we actually construct a plan, we can see that it emerges from a deliberative process—from a back and forth between thinking about the goal and then shifting to the various steps it will take to reach it. This is what the self-reflective mind is all about—a constant dialogue between the big picture—the One— and the specifics of which it consists—the Many.

The large size of the great ape brain and the emergence of the PFC gave rise to a self-reflective intelligence, and this opened the door to new, more complex kinds of behavior. While I doubt there is any single, neurological center in the brain to which we might trace the presence of a "self," the evidence suggests that the prefrontal cortex is an essential part of the neurological infrastructure necessary for the emergence of self-reflective capacity.

Conclusion

Human evolution could only have happened at a certain point in the evolution of the animal world. It was made possible by the emergence of a new cognitive threshold among some of the higher vertebrates. As a result of certain changes in the brains of these animals, they were able to construct more complex models and perform more complex behaviors. However, only the great apes, with their unique combination of anatomical features, were in a position to take full advantage of the behavioral options opened up by this new cognitive threshold. As they began to rely on these new behaviors, they set a unique evolutionary process in motion.

The intelligence of the great apes, then, occupies a special place in the evolutionary hypothesis presented in this book. It represents a transitional stage between an earlier, less developed level of intelligence and a later, more highly developed one—that which is associated with our early hominin ancestors and which was capable of creating a new, culture-based way of life. Great ape intelligence, then, occupies the first rung on the ladder between animal and human intelligence.

Summary of the Argument

In chapter three, I examined tool using and raised the issue of "indirect" ways of solving problems. Such behavior is too complex, that is, involves too many discontinuous steps, to be achieved through trial and error methods. An animal must work out the solution—put the pieces together in its head before it would be able to join them in practice. The fact that the great apes can solve these kinds of complex problems in their heads suggests they possess a capacity that has something in common with human reasoning.

To understand reasoning, I turned to Western philosophy. Reasoning, this school concluded, is based on an ability to represent objects and processes as systems: a capacity to build "logical" models of things. The ability to model things as systems enables a being to see how things are related to one another—object-object relations—and to grasp cause and effect. This enables a being to predict how affecting one part of a system will affect the other parts. This is the foundation of reasoning—and it is clear that the great apes possess the rudiments of such a capacity.

That answer, however, posed another question. What is distinct about the brains of great apes that enables them to model things differently than other animals? At this point, I looked at evidence that argued that all vertebrate brains share a common design—all are built to organize experience into systematic models. What, then, distinguishes great apes? What allows them to solve problems in their heads that other animals can't?

To get at this, I turned to the work of Antonio Damasio. The key to a higher intelligence lies in the emergence of another center of awareness. This second center enables the mind to monitor its own processes—its own thoughts. This dual awareness gives rise to a self-reflective intelligence—an intelligence that is based on the mind's ability to re-process its own models—think about its thoughts. A mind that is aware of itself gives rise to a being that can think about itself as an "object"—about how it appears in the eyes of others. This allows it to see itself as a part of a larger whole—a critical development that opened up unprecedented social capacities. This is the origin of self-hood. I then examined evidence that supports the conclusion that the great apes are capable of a degree of self-reflection.

Lastly, I looked for evidence of some neurological feature that might distinguish great ape brains from those of other animals—something they might have in common with humans. The answer seems clear; both the great apes and humans share a uniquely developed pre-frontal cortex. The PFC plays a central role in planning and reasoning.

Seeds of a Higher Intelligence

Great ape cognition contains the seeds of a higher intelligence. While it is able to produce simple forms of tool using, it is not at a level where it can generate a cultural survival strategy. Great apes are not able to communicate in the same way we do, nor are they capable of organized, cooperative behavior.

Based on current information, it appears that it may have taken on the order of three hundred thousand years for our australopith ancestors to have evolved a brain that could produce effective stone tools and some level of cooperative behavior. That's the time between the onset of the East African ecological crisis (2.9 million years ago), and the stone tools left in the carcasses of large herbivores (2.6 million years ago). If these meat-eating hominins were a form of *Homo habilis*, we can assume that their brains were 35% larger than those of the australopiths. It would be another couple of hundred thousand years before that brain could move up another notch and provide these early hominins with the cultural capacities that would allow them to come down from the trees and reside permanently on the ground.

While a great ape level of intelligence could not produce a cultural way of life, it could start the ball rolling—tool using. It was, I believe, the essential ingredient that made it possible for our lineage to get on a cultural evolutionary track. While great ape intelligence is clearly far less developed than our own, nevertheless, the question arises as to whether it should be looked at as a simpler version of our own kind of intelligence. Is our intelligence on a continuum with theirs?

To be able to solve the riddle of human evolution, we have to set aside categorical thinking—the idea that something must be either one thing or another. In some ways, our intelligence is on a continuum with that of the great apes, and in others, it represents a leap to another plane. Their brains were able to get the human evolutionary process started, and for most of human evolution, it appears that the brains of our archaic ancestors were basically up-scaled versions of theirs. At the very end of our species' evolutionary process, however, a change took place in the shape of the brain of our *Homo sapiens* ancestors. This is when a fully-developed language capacity emerged. It was language—a cultural tool— that carried our vertebrate intelligence onto another plane.

Our great ape ancestors, then, possessed an intelligence that made culture possible. We, on the other hand, possess an intelligence that culture made possible. But this latter development lies ahead of us. At this point we need to return to our evolutionary narrative.

Chapter 7

Coming to The Ground: *Homo erectus*

Before our excursion into tool using and intelligence, we had arrived at a point where we began to find something new—stone tools in the carcasses of large grazing animals and the presence of large-brained, australopith-like creatures. While no fossil hominin remains have been found in direct association with these early stone tools, the appearance of larger-brained types at around the same time would seem to be no accident.

As simple and crude as the first stone flakes were, they lay beyond the capacities of any contemporary ape. In an interesting experiment, Toth and Schick (see chapter 3) attempted to teach a chimpanzee to make the same kinds of simple stone tools manufactured by our early ancestors. They worked with "Kanzi," a very intelligent bonobo (pigmy chimp) who had excelled in learning to communicate with signs. After months of frustrating effort, they concluded he just wasn't up to the task: "Although Kanzi is still continuing to improve his tool-making abilities, his present level of expertise is significantly below that seen in Oldowan hominids (*H. habilis*). ... He doesn't show the understanding of flaking angles that the Oldowan hominids had" (Toth and Schick 1993, p. 139).

Recently, a new and more comprehensive study of chimpanzee tool using capacities elaborated on Toth and Schick's observations (Bril 2012). While confirming their view that chimps lacked an ability to recognize the proper striking angle, the research brought to light another dimension of the problem. To be able to deliver a blow at a proper angle, the toolmaker has to coordinate the movement and positioning of both hands. This seemed to be the major obstacle for chimps: too much to process and coordinate.

One aspect in the evolution of the human brain that is often overlooked is the extraordinary size of the cerebellum. We tend to think that the expansion of our brain was all about the cortex—the area where the highest cognitive tasks are performed. However, the human cerebellum, which plays a major role in our capacity to learn and execute complex motor activities is every bit as large, proportionally, as are our cortical regions.

Homo habilis

The idea that stone tools and larger brains are connected goes back to the dawn of modern paleoanthropology—to the work of Mary and Louis Leakey. Working during the 1950's in Olduvai Gorge in Tanzania, the Leakeys uncovered a number of very primitive stone implements. Certain that the makers of the tools must lie buried nearby, they spent years in the Gorge searching for their remains.

In 1959, they discovered the skull of a kind of australopith. However, its primitive appearance and ape-sized brain made it an unlikely candidate for the creator of the stone tools. The following year, as chance would have it, they unearthed a different, smaller, but larger brained creature. This hominid had a brain that they originally estimated at 700cc, as opposed to the 450cc brain of the australopiths. (We now know that the actual size of its brain was even larger—800cc.) While this new fossil was clearly very closely related to an australopith, because of its larger brain, the Leakeys concluded that it must have been the toolmaker they were looking for. The combination of large brains and stone tools led the Leakeys to put this australopith-looking fossil into the human genus. They called it *Homo habilis*—"handy man."

The Leakeys dated the *H. habilis* fossil to 1.75 million years ago—a figure that has been only slightly revised as the result new dating technologies (to 1.8 mya). Based on what was known at the time, that seemed to give it a sufficient time depth to make it a plausible ancestor of the human lineage. However, a number of more recent discoveries began to cloud the picture.

A different looking, older (1.9 mya) large-brained *Australopithecus*-like creature was discovered in northern Kenya—*Homo rudolfensis*. Then a very advanced fossil form—now classified as a *Homo erectus*—was discovered and dated at 1.8 million years. *H. erectus*, as we shall discuss shortly, is very different from both *H. habilis* and *H. rudolfensis*. Those discoveries were followed by fossil discoveries made in the Republic of Georgia and also dated at 1.8 mya. The Georgian hominins demonstrate a confusing blend of small brains (600cc and less), advanced looking faces, and a clear association with simple stone tools.

The more we learned, the more complicated the picture became.

An Important discovery

When the Leakeys discovered *Homo habilis*, it was believed they had found the "missing link"—the intermediary between *Australopithecus* and the first fully terrestrial, large-brained human type—*Homo erectus*. Now, however, with the discovery of a number of different early, large-brained hominins dating from more or less the same time period—some of whom were even outside of Africa— it became difficult to conceptualize a simple path from great ape to early human. It was becoming clear that there had been more than one australopith species, and that some of these may have been evolving in the same general direction.

The main problem in sorting this out was a total lack of fossil evidence from the critical period between 3 and 2.3 million years ago. This is when the use of stone tools and meat-eating became widespread, and based on the fossils found at the end of this period, this is when the transition from great ape to some form of hominin took place. However, the absence of fossil evidence created a serious gap in our understanding of the dynamics of this process.

Then, in 2015, a team of paleoanthropologists from Arizona State University reported the discovery of a partial lower jawbone dated to 2.8 million years ago (Villmoare 2015). The jawbone was found in the Afar region of Ethiopia, an area where the remains of over a 100 different *Australopithecus afarensis* fossils (Lucy types) had been unearthed, and where 2.6 million-year old stone tools had been found. The partial mandible, which still had most of the teeth in place, demonstrated a mixture of australopith and early *Homo* characteristics.

The ASU team also published a report describing the climatological and ecological environment in which the jaw had been embedded (DiMaggio et al. 2015). East Africa, as noted earlier, had experienced a dramatic climate change between 2.9 and 2.6 million years ago. The ASU team was able to document the transformation of the Afar region from a moist, woodland environment to a drier, open savannah—something like what the Serengeti is today. The australopiths living in this region would have been under tremendous pressure to find some way to adapt to these changing conditions.

In the same month, another research team reporting in *Nature* (Spoor 2015) presented the results of an in-depth analysis of the jawbone and skull of the original *H. habilis* discovered by the Leakeys.

They arrived at two important conclusions. The first is that the brain of *H. habilis* was actually larger than originally estimated: 800cc as opposed to 700cc. And the second is that the jaw is very primitive—closely resembling that of an *A. afarensis*. It is more primitive than that of *H. rudolfensis* or any other early *Homo* yet discovered in Africa.

These findings would seem to suggest that *Homo habilis* may very well have been the missing link it was originally thought to be. This data indicates that *A. afarensis* may evolved into *H. habilis* sometime between 2.8 and 2.5 million years ago, and that as *H. habilis* spread out, it served as the rootstock from which a variety of later hominin types evolved.

However, all the early hominins (*habilis*, *rudolfensis*, and the Georgian variety) are quite problematic. In the African cases, they had very large, but also extremely variable brains—ranging in size from 550cc to as high as 800cc. This extreme range of brain sizes suggests a high degree of sexual dimorphism—that males were much larger than females.

Taken together, the brains of these early hominins averaged around 35% larger than those of

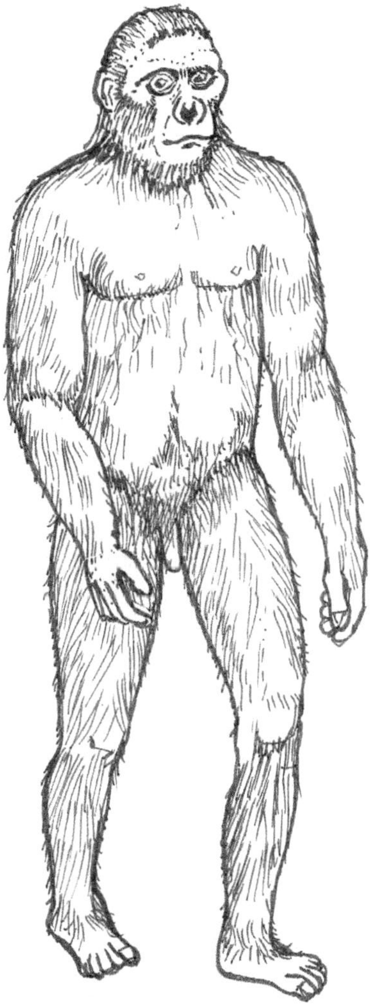

Homo habilis

the australopiths. This implies an immense behavioral difference. The australopiths, we must remember, had very large brains. Relative to their body size, their brains were larger than those of any animal that had lived before them. And the brains of these hominins were almost half again as large as theirs. This leap in intelligence was undoubtedly about more than just stone tools; tools were almost certainly just one element in a complex of newly evolving capacities.

In spite of their dramatically bigger brains, these early hominins were a lot like their australopith ancestors. They were small (less than 5 feet tall), and most importantly, they had not yet severed their ancient anthropoid connection to the trees. Based on what little evidence we have of their skeletal remains, it appears that much of their anatomy was still adapted for an arboreal way of life. Their arms were longer than their legs, their fingers were long and curved, and their shoulders and torso retained their basic australopith design.

Homo erectus

Sometime around 1.9 million year ago, a new kind of hominin emerged—*Homo erectus*. There is no debate about how to categorize this creature: it was not some kind of great ape—it was clearly one of us.

Once again, we see a correlation between an evolutionary development and dramatic climate change. Beginning around 1.9 million years ago, eastern and southern Africa experienced another pulse of dry weather—and again the forests and woodlands retreated and grasslands and desert expanded to fill the void (deMenocal 2011, p. 540). It was no accident that this would be the moment we would see the emergence of the first member of our lineage that would finally abandon its ancient dependence on life in the trees. *Homo erectus* is the first hominin that is fully committed to living on the ground.

Homo habilis and other early hominins were really halfway creatures. The lower body they inherited from the australopiths enabled them to move relatively efficiently on the ground. Their upper body, however, was still designed largely to meet the demands of arboreal life. This limited the degree to which their hands and arms could be re-shaped for tool using. It also interfered with the re-shaping their lower bodies for maximum bi-pedal efficiency.

The future of the hominins clearly lay on the ground. That's where their food was located, and that would enable them to evolve an anatomy that was better suited for cultural skills. However, coming to the ground was no simple matter. The African savannah was an extremely dangerous place—home to the world's fiercest predators. Two and a half million years ago, the plains were occupied by large numbers of huge, archaic grazing mammals (megafauna) that were hunted by huge, archaic predators—saber-tooth cats, giant hyenas and the like. Small-bodied ape-like creatures, lacking claws and canines and armed only with simple stone flakes and sharpened tree limbs would have been no match for these animals.

The early arboreal hominins, it would seem, were caught in something of an evolutionary conundrum. They had to develop better cultural skills before they could come to the ground; but they couldn't develop those skills until they came to the ground. They were, as we like to say, between a rock and a hard place. The evidence suggests it took a few hundred thousand years to solve this chicken and egg dilemma. The solution was *Homo erectus*.

From the neck down, *H. erectus* was fundamentally the same as us. It was about as tall as contemporary humans, the males reaching 5'10" or more, and the proportions of the body were the same as ours; their legs were longer than their arms. While no remains of a hand have yet been found, the shape of the torso, shoulders and arms all indicate they were no longer built to climb like an australopith.

In general, there has been a tendency to think about our species' descent from the trees in terms of freeing

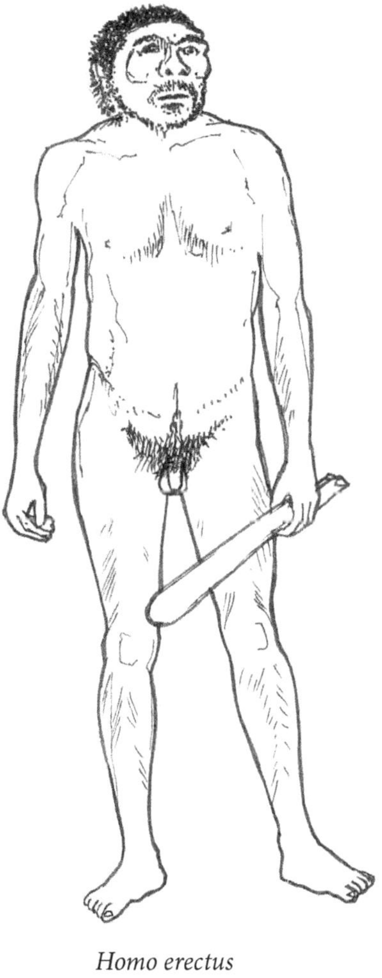

Homo erectus

the hands for tool using. And, of course, that was vitally important. However, freeing the hands was inseparably connected to freeing the arms. The hands of apes are a part of a climbing complex: an integral scaffolding in which hands, arms, shoulder joint, back and chest muscles are all woven together. The arms of great apes are not as flexible and free-swinging as ours: they are anchored in the powerful musculature of their torsos. Apes climb and swing with their whole upper body. That is what makes them so strong and enables them to move about so effortlessly in the trees. Coming to the ground didn't just free the hands—it also freed the arms.

The importance of this is not generally appreciated. While many have pointed to the fact that changes in the arm and shoulder joint enabled the arms to swing freely during running—a key factor in balancing the body—its significance goes way beyond that. In a letter to the magazine *Nature* (Roach et al., 26 June 2013), a team of researchers reported the results of experiments on "high-speed throwing" capacities in human beings. They concluded that the ability of humans to throw projectiles rapidly and accurately depends on a series of anatomical changes that took place as the arms were released from the requirement of bearing the full weight of the body during climbing movements: "These features first appear together approximately 2 million years ago in the species *Homo erectus*" (ibid., p. 483).

Homo erectus was the first member of our lineage that could effectively throw a stone and/or a spear. It could deliver lethal force at a distance—a valuable attribute for a predator: "High-speed throwing was probably a critical component of a suite of hunting behaviors that enabled early members of the genus *Homo* to thrive in new and varied habitats both in and out of Africa" (ibid., p. 486).

Bipedalism

The changes that took place in the lower body were every bit as extensive as those in the upper anatomy. While the australopiths had solved certain critical engineering problems posed by bipedalism, *Homo erectus* made a number of important improvements. Its legs are much longer—which would have given it a long striding gait for walking and most importantly, a capacity for rapid and extended running.

Everything about *H. erectus* indicates a being that is built for strenuous movement on the ground. Its long limbs and tall, lithe body are well designed for dissipating heat. The chest is barrel shaped, deeper and more rounded than that of its more ape-like predecessors, and thus able to accommodate larger lungs. It had evolved our diaphragm-based deep breathing anatomy and appears to have been particularly well designed for running.

There is currently a good deal of speculation that *H. erectus* may have occupied a special niche among the predators of the savanna. The hypothesis is that it was a daytime hunter, running animals down in the heat of the sun—while leaving the night shift to the saber-toothed cats, hyenas, leopards, etc. Contemporary humans have the most highly developed biological cooling system in the animal world. We have lost our great ape fur coats and have more sweat glands per square inch than any other animal. If *H. erectu*s did indeed possess all these features—lack of fur, enhanced sweating capacity, long, lithe body, deep breathing system—it would have had an exceptional potential for long distance running under hot conditions. This is something most other animals are ill equipped for.

Initially, I found the idea that these early humans could have run down much faster four-legged animals highly improbable. Then I stumbled upon a diary of an American colonist who became caught up in the French-Indian wars and was held captive by a branch of the Mohawks for five years. In it, he relates how these Native Americans told him they could run down every animal "except the wolf" (Drimmer 1985). When would our species have developed this capacity? While it is not possible to provide a definite answer at this point, it is entirely plausible that *Homo erectus* was capable of this behavior.

Acheulean Stone Tools

Homo erectus had a skull—a face and brain case—that looked a lot more like our own than those of earlier hominins. However, the brain capacity of the first *H. erectus* to come on the scene (sometimes called *Homo ergaster*) was only slightly larger than that of its more australopith-like predecessors—850cc. While the changes in the arms and hands undoubtedly provided them with improved tool-making and tool-using skills, their advent did not immediately signal a revolution in stone technology. It appears that the earliest *H. erectus* used the same basic stone tool making techniques (Oldowan) as *H. habilis* and other early hominins.

However, by 1.75 mya, we find the presence of a radically new stone technology. The earlier, Oldowan stone industry was based upon striking two stones together. One of these, the hammer stone, was used to knock flakes off the other—the core. The flakes, with their sharp edges, were used to cut hides, meat, wood, or whatever. What was left—the core—was used to crush bones for marrow. This technology was not as simple as it looks and took a good deal of skill to produce. However, the new stone industry—called Acheulean—was much more complicated and required a higher skill level.

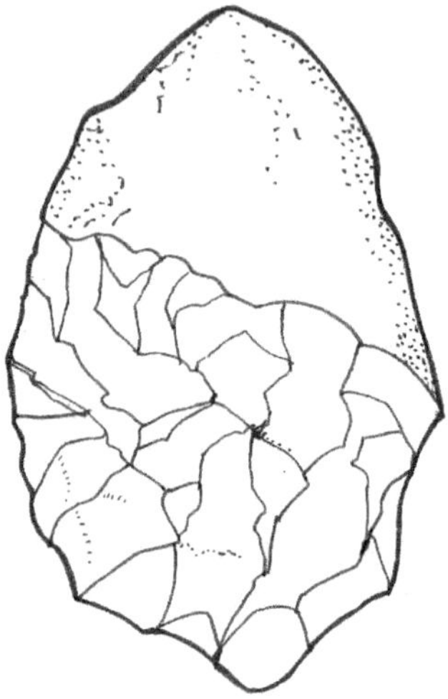

With Acheulean tools, it is not the flakes, but the core that is the main goal of the manufacturing process. Instead of starting with a river pebble, the toolmakers began with a large stone blank. They then knocked flakes off of the top and the bottom, whittling the blank into a large, sharp-edged tool that paleoanthropologists call a "hand-axe." These tools involved a good deal of planning and foresight and demonstrate a uniformity of size and design that had not been found previously. They had a standardized "pear" or "tear-drop" shape and are the earliest examples of a capacity to create an object on the basis of a mental image of the desired end.

The Emergence of a New Way of Life

There are many unanswered questions about *Homo erectus.* While it was clearly related to *H. habilis* and *H. rudolfensis*, it was quite different from both of them. Most likely there is an intermediate form that has not yet been discovered. It almost certainly evolved somewhere in eastern or southern Africa—but the discovery of early, more primitive hominins in the Republic of Georgia has raised an intriguing possibility. It is now clear that *H. habilis*-type hominins left Africa quite early, and it is possible that one of these could have evolved into *H. erectus*, which then migrated back. Very unlikely, but given the gap in the data, not impossible.

From our perspective, the most important fact about *H. erectus* is that it was the first human that had fully abandoned arboreal life and that was able to hold its own on the ground. These hominins were not restricted to scavenging or opportunistic predation: they were hunters—capable of pursuing prey on the open savannah.

H. erectus, as the Georgian finds indicate, was not the first hominin to leave Africa. However, once it appeared, it rapidly spread throughout the temperate regions of the Old World. By 1.2 million years ago it had reached China and other parts of Asia, and shortly thereafter could be found in southern Europe. It appears to have been a relatively successful species, and by a million years ago, was the only branch of our lineage still standing. The various australopiths and all their *H. habilis* descendants—except the Flores "hobbit"—had disappeared.

Homo erectus occupies a key place in the emergence of the human lineage. It represented the final step in the transition from arboreal to terrestrial life. After it, the human body below the neck changed very little. The only significant later skeletal developments involved changes in the female pelvis associated with delivering larger-brained fetuses.

The adaptation to life on the ground, however, was not simply about changes in anatomy. *H. erectus* could only have survived on the ground if it could have developed behavioral capacities that would have allowed it to compete with and defend itself from powerful predators. It needed to develop a whole complex of cultural capacities. It was no easy matter for our ancestors to come to the ground and adopt a more predatory lifestyle. It would have involved the development of a level of cooperation that is not found in any of the existing great apes.

Large brains are associated with longer pregnancies and an extended period of infant dependency and childhood. Large-brained offspring are a great burden for the female of a species. The scant remains that have been found of *Homo erectus* females indicate that they had much wider hips than the males and were a good deal shorter. It is hard to imagine how they could have hunted during certain critical periods of pregnancy and childrearing. The implication is that if hunting was to play a significant role in the lives of these early humans, they must have evolved some kind of division of labor: some way in which the males provided meat for the larger community. This was not a simple matter; food sharing, as we shall discuss in a later section, is not something that comes easy to our great ape cousins.

The second important adjustment would have been collective hunting and defense. It is one thing to slay an animal and another to prevent other carnivores from taking the carcass away. Recent research has revealed that the human tapeworm evolved from the same species as that which inhabits the intestines of wild dogs and hyenas. The implication is that our early ancestors were able to drive these animals off of their kills—and in the process, acquired the parasites that were present in their saliva.

Intelligence

By 1.5 million years ago, the early variety of *H. erectus* (*H. ergaster*) had evolved into a larger brained kind. This latter's brain measured 970 cc—fully twice the size of an *Australopithecus*. It made better tools (Acheulean) than its predecessors, and there are claims that it may have been able to make use of fire.

At this stage, *Homo erectus* appears to have reached a relatively stable biological equilibrium: one that may have lasted as long as 700,000 years. It was undoubtedly quite intelligent—having a brain that was twice as large, relative to its body weight, as any non-human animal living today. Its intelligence, however, is only partly revealed by its skillful tool manufacturing. The key to its survival would have lain in the development of new social capacities. And these capacities, like those that distinguish humans at every stage of their evolutionary development, depended on their ability to make use of objects that were not built into their biology.

Chapter 8

Sharing Information

As our early ancestors began to abandon their arboreal refuge and descend to the floor of the savanna, they would need more than stone flakes and sharpened branches to hold their own in this new environment. Tools, of course, were absolutely critical to their adaptive process. The biology they inherited from their great ape lineage was not well equipped to survive in this new environment. At the same time, however, tools alone would not be sufficient to ensure their survival.

The tools these early hominins possessed were very primitive. It would be more than a million years before their descendants would be able to haft a stone point onto a wooden shaft—to make a truly effective spear or axe. They did not yet have spear throwers (atlatls)—means of extending their arms and adding heft and distance to the throw. There were no bows and arrows. Indeed, in the early stages they even lacked fire. What implements they did possess enabled them to dig up tubers and root crops, kill small and moderately sized animals, crush bones, and cut through hides and sinew. But there was no way that they could carve out a place for themselves on the savanna with the limited capacities these tools provided.

What other resources were available to them? As was the case with tools, they would need to tap into their great ape intelligence; and as with tools, they would figure out how to generate capacities that went beyond the bounds of their biological inheritance. Indeed, they would need to create another kind of "tool."

Besides raw nature, what other resource could they use to enhance their innate capacities? They had their relatives and neighbors—their social community. To be able to fully abandon life in the trees, the early hominins would have to develop the collective potential of their ancestral great ape troop. If they were to survive in their new circumstances, they would have to transform an anarchistic, biologically driven social network into a community that would be able to organize itself in unprecedented ways.

Their newly emerging way of life required a social base that could generate force levels necessary for hunting larger animals, defending the kill, protecting the group, and addressing the problems posed by providing help for vulnerable offspring and their mothers. This would involve food sharing. Indeed, on some level, males would be required to take on some of the burdens created by large-brained offspring. The early hominins would have to develop a capacity for cooperation that went beyond anything that existed among the great apes—indeed, beyond that of any other animal. That would be no simple matter.

There were a number of different dimensions to the development of human cooperative capacities. Some involved emotional developments—most importantly, impulse control, particularly aggression. Much of this depended on developments in the pre-frontal cortex. But more than simply restraint, social life is also built on positive incentives—desire for companionships, protection, sex, etc. This may well have been the time when important changes took place in the human sexual response: when periodic estrus episodes were replaced by perpetual sexual receptivity. This would have played—as it does today—a powerful role in creating constant and enduring male-female bonds.

However, of all the developments that went into producing new social capacities, none was more important than the emergence of a new form of communication. This was the decisive step in the transformation of the great ape troop into a different kind of community.

Organization and Communication

There are a number of animals that are capable of spontaneously coordinating their efforts and coming to each other's assistance. Social carnivores, lions and wolves in particular, are adept at this. Elephants and musk oxen form a circle around the young when predators threaten the herd, but perhaps no animals are more impressive than porpoises. Guided by their echolocation systems, porpoises are able to perform amazing feats of coordinated behavior. In one example, a pod of porpoises forms an ever-tightening circle around a school of fish, constricting them to such a degree that the individual fish have no room to maneuver and can be consumed at will.

In all these instances—and there are many others—a collective effort takes shape spontaneously, without any exchange of information about group goals or individual roles. It's clear that some animals — though not as many as generally believed—can pick up cues from one another and adjust their behaviors to accomplish certain limited collective ends. We see this with chimpanzees when they position themselves to surround a fleeing monkey or move silently in single file through the forest to ambush another troop. All these examples take place without any ability to communicate aims or means. Accordingly, when the chimps arrive at the battle scene, it is pure chaos—there is no plan, no coordination of any kind.

Communication is the limiting factor in cooperation. In order for numbers of individuals to be able to coordinate their efforts, they need to share certain kinds of information. To begin with, all the actors need to know what the goal is—what it is they are collectively trying to accomplish. Secondly, they need to share a plan—some kind of framework that enables different individuals to understand how their actions fit together with those of others to achieve the goal. For different individuals to be able to coordinate their behaviors, they have to be on the "same page." The development of a capacity to share information of this nature was a turning point in human evolution: it gave birth to a new power—organized social action.

Organization, like technology, endows a group with capacities that are not part of its biological inheritance. It enables the energy of different individuals to be combined in a way that produces a force that is greater than the sum of those individuals acting alone. Coordination enables tasks to be accomplished that are not feasible without it. Ten individuals working separately and serially cannot move a heavy log. By combining their strength and coordinating their efforts, however, an organized group of individuals can accomplish what an infinite number of isolated individuals could not. In the same vein, individuals working alone cannot drive animals into an ambush. The ability to synchronize the efforts of individuals that are not in immediate face-to-face contact—separated by both space and time—was a huge plus for our evolving species. However, in order to accomplish this, they would have to invent a new way of sharing information.

The emergence of a specifically human communication capacity has been, perhaps, the least understood aspect of human evolution. The problem is that our communication system has been looked at as a biological capacity. That's part of the picture. However, in order to exchange information about the external world, a being needs a special kind of "tool." A human form of communication, then, is not a biological, but rather, a cultural behavior.

A Different Principle

Human communication is based on a different principle than animal communication. This is not generally understood. Indeed, the whole area of communication has been a major source of confusion and misunderstanding in evolutionary anthropology. While it is not fair to blame sociobiology for every misconception about the nature of the human way of life, their biologistic framework has not helped matters. The assumption that there are no "fundamental" differences between other animals and humans produces a research strategy that looks to other creatures for the roots of human behaviors. Our capacities are seen as simply an extension of theirs. This misses the mark entirely.

Human behavior emerged precisely at the point where our ancestors began to do something different than what other animals were doing. They learned how to make use of the properties of things that lay outside their organisms. While it is easy to recognize this principle at work with stone tools and other kinds of technologies, it is not so obvious in the case of communication. The fact is, however, that our ancestors used this same formula in developing a new way of exchanging information.

Neither humans nor any other creature is born with a built-in capacity to transmit information about the external world. To be able to do that, a being needs to create a special tool.

In order to clarify this whole issue—the difficulty in transmitting information about the external world and how our early ancestors overcame it—I would like to take a brief look at how other animals communicate.

Expressive Communication

Like everything else we are learning about animals, their forms of communication are much richer and more complicated than was once believed. The signals they use, like ours, transmit important information. However, the nature of that information is very limited. The vast majority of their signals are spontaneous and expressive. Most arise from impulses that are built into an animal's biology and that inform others about its different moods and internal states: what's going on inside of it (sexual receptivity, pain, hunger, etc.) and how it is reacting to events in the world around it (fear, excitement, caution, etc.). These signals do not, however, enable animals to designate or describe the nature of external objects or events.

Animal signals consist of a variety of vocalizations, facial expressions, postures, movements of different body parts (ears, tails, etc.), and various chemical secretions—activities that more or less automatically resonate with other animals. A wildebeest, spotting an approaching lion is seized by fear. The sounds it emits, the posture of its body, the position of its tail and ears, even the odor of the nervous sweat it exudes, serve to alert other members of the herd that danger is present. Others of the same species—and indeed members of many unrelated species—can read these signals and respond more or less automatically to them.

This "expressive" form of communication, depending on the intelligence of the animals involved, can be quite complex and nuanced. While it would seem to be very limited—an animal can only inform another about how it is responding to events—in fact, the situation is not quite that cut and dried. An animal's internal response has implications about what it is responding to. While animal communication is primarily based upon the ability of animals to automatically sense each other's emotional or internal states, it can have a volitional and intellectual dimension as well.

Crows, for instance, have "alert" calls that seem less like spontaneous emotional responses than intentional efforts to inform others about the presence of danger. They are even able to modulate the intensity of these calls to reflect the degree of threat posed by an intruder. Nevertheless, these calls are incapable of specifying the nature, number, or location of the threat. The signals used by animals to communicate thus lack a "referential" capacity—an ability to refer to specific objects or events in the external world.

Having referential implications is not the same as being a referential signal. If a wildebeest wishes to specify a particular object like a lion, for instance, it cannot do so with a signal that can only carry emotional information, i.e., fear. It is also afraid of leopards, elephants, rhinos, hyenas, cobras, and wild dogs. If a crow wanted to alert others that the danger comes from a human hunter rather than a cat, it would have to be able to use a signal that could indicate that difference. Neither a wildebeest nor a crow is able to generate such a signal.

Vervet Monkeys and Honeybees

With the ascendancy of sociobiology, the distinction between referential and expressive communication came under increasing criticism—and, indeed, has largely fallen into disuse. Honeybees, it was pointed out, are able to communicate specific information about their environment. The "waggle dance" enables them to inform other members of the hive about the direction and distance of pollen sources.

Even more telling is the fact that vervet monkeys make three different warning calls which alert other troop members whether danger comes from the air, the ground, or the trees—from eagles, leopards, or snakes. These calls, it is claimed, are not expressions of inner emotional states, but rather refer to specific objects in the outside world. We now know that Diana monkeys are able to communicate similar distinctions.

On the basis of these few examples—which, it should be mentioned, were not subjected to critical analysis—the distinction between referential and expressive, that is, between human and animal forms of communication was believed to have been discredited. Indeed, the use of those terms has been largely abandoned. This was a serious mistake. The distinction between expressive and referential communication is real and significant. One relies on a signal that is spontaneous, that is, automatically generated by the organism. The other makes use of one that must be deliberately constructed.

Animals are biologically built to

resonate with each other's emotional states—to sense what another is feeling and expressing. They cannot, however, know what another is thinking—the information it possesses about events in the external world. There is no automatic or built-in way that any being can make its perceptions and thoughts known to another. Years of research into ESP have failed to find any evidence for believing that this might be possible. To be able to share information about the external world, a being needs a tool: a signal that refers not to what it is feeling, but rather, what it is thinking.

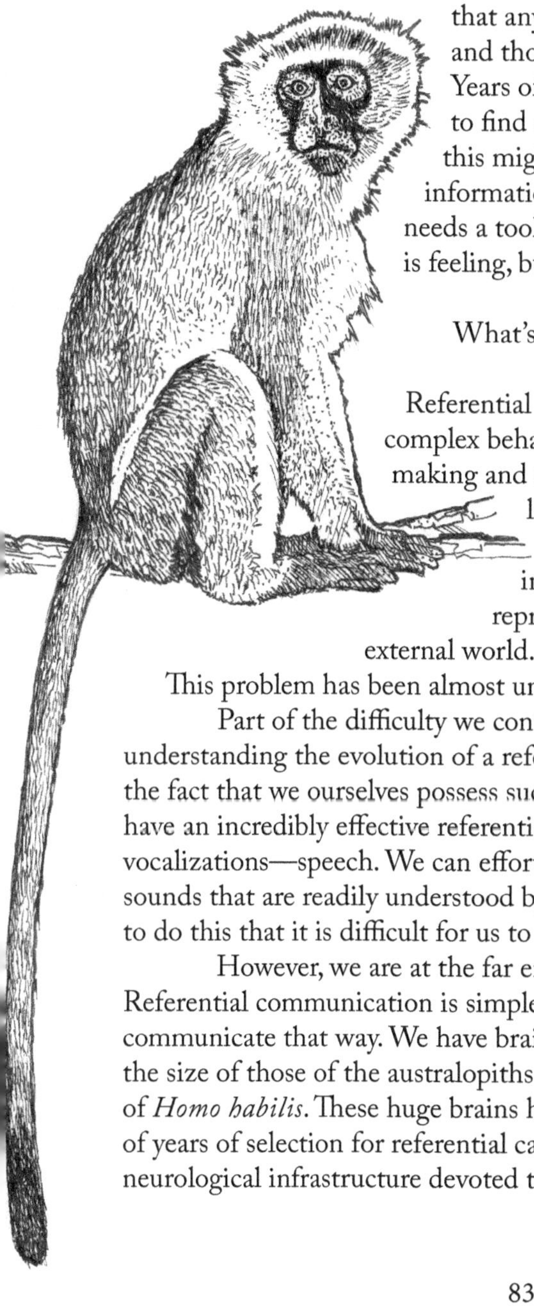

What's So Difficult About Referential?

Referential communication is an exceedingly complex behavior, much more complex than making and using stone tools. The problem lies in the fact that the signal—the "tool" used to transmit information—must be able to represent an object or event in the external world. The key word here is represent. This problem has been almost universally overlooked.

Part of the difficulty we contemporary humans face in understanding the evolution of a referential capacity derives from the fact that we ourselves possess such an aptitude: language. We have an incredibly effective referential system built on the use of vocalizations—speech. We can effortlessly translate our thoughts into sounds that are readily understood by others. Indeed, it is so easy for us to do this that it is difficult for us to imagine why other animals can't.

However, we are at the far end of a long evolutionary process. Referential communication is simple for us because we are built to communicate that way. We have brains that have grown to three times the size of those of the australopiths and are twice as large as those of *Homo habilis*. These huge brains have been shaped by millions of years of selection for referential capacities. They possess an entire neurological infrastructure devoted to encoding experience into signs and sounds and processing the messages we receive from others. Without that infrastructure, we would not be able to extract meaning from the sound signals we receive.

The complex of elements that enable us to communicate with language is not something we are automatically aware of. We tend to assume that when our ancestors began to develop a new way of communicating, they must have started with some simplified form of speech. A popular hypothesis is that the process began when our early ancestors recognized that vocalizations—their great ape calls—could be used to represent certain things they wanted to refer to. Little by little, the story goes, they acquired the capacity to make and remember more sounds (words), eventually developing certain rules for stringing these sounds together into narratives (grammar) and voilà—they had language. That's almost certainly not the way our language capacity evolved.

In that model, referential communication and language are treated as synonymous. The accumulation of language skills is portrayed as the path to a referential capacity. However, the evidence suggests that it happened the other way around. Our ancestors first developed a referential capacity—a way of exchanging information about the external world—and that system provided the foundation upon which language later evolved.

Not Based on Sound

The most difficult hurdle in understanding how our ancestors went about creating a new way of communicating, is letting go of the idea that it must have been based on vocalizations. There are two arguments in particular that would seem to rule that possibility out.

The first is that great apes do not have voluntary control over their vocal apparatus. This realization came as quite a surprise. While we have no direct data about the australopiths, we now know that the living great apes cannot voluntarily control their vocalizations. Their vocal apparatus is neurologically wired differently than ours (Deacon 1997); their vocal chords are directly connected to the emotive centers in the brain, and, as a result, "the production of a sound in the absence of the appropriate emotional state seems to be an almost impossible task for a chimpanzee" (Goodall 1986, p. 125). Chimp calls are biologically wired to be expressive.

This was not understood when the first attempts were made to teach chimps human language. Efforts to get them to "speak" invariably produced disappointing results. The first attempt was made during the 1930s when Winthrop and Luella Kellogg raised a chimp baby along with their infant son. The chimp learned to understand about 100 words, but never attempted to speak. Keith and Catherine Hayes undertook an even more intense effort in the 1940s. After years of training, they could get "Vicki" to say only three words: mama, papa, and cup.

In 1967, Allen and Beatrice Gardener decided to try a different approach. They switched from vocalizations to hand signals, teaching the baby chimp "Washoe" elements of American Sign Language. As a result, Washoe was eventually able to learn how to use some 350 signs to represent both specific objects and actions. Observations of chimps, both in their natural habitats as well as in experimental situations, reinforce the view that the seeds of a referential capacity are not to be found in their vocalizations, but rather in their rich repertoire of gestures and facial expressions (Tomasello 2008).

The second argument against the idea that the earliest forms of referential communication were based on sound is the fact that sound is actually a poor vehicle for representing objects and events in the external world. That statement seems to fly in the face of our everyday experience. For us, sounds—in the form of words—are extremely effective tools for representing or referring to things. We assimilate them easily, have no trouble remembering them, and experience them as inseparably connected to the things they refer to.

However, when we take a closer look at these sounds, we can begin to understand that they could never perform the role they play for us if we were not biologically constructed to make use of them. The fact is that the sounds we use have no necessary connection to the things they refer to. The relationship of a particular sound to some object is entirely arbitrary. We could use any sound—dog (English,) perro (Spanish), canis (Latin), esquintle (Nahuatl), gou (Chinese), inu (Japanese), kopek (Turkish), sobaca (Russian), kalb (Arabic), etc., etc., to refer to that friendly little creature wagging its tail over there.

As counterintuitive as it may seem to us, sound is not a good material for representing objects. Most of the things we represent with sounds don't make sounds— rocks, sky, snow, clouds, spears, tools, water, seeds, fruit, wood, meat, go, come, here, there, after, near, far, etc.—and have no sound equivalent. The sounds we use are "empty" —they contain no information about the objects they signify. The fact that these sounds are empty, that their forms have no relationship to their meanings, is what defines them as "symbols." Their meanings are established purely by convention, that is, by the consensus of a community.

Humans ended up using sounds for referential purposes, but that is not where they began. To be able to use sounds as symbols, that is, as signals that could stand for objects, was beyond the capacity of our early hominin ancestors. To communicate referentially they would have had to find a signal that was consonant with their level of development. What would that have been?

Finding a Referential Signal

If it is highly improbable that our species' referential capacity evolved directly out of anthropoid vocalizations, what other option is there?

There is something that can be used as a referential signal that is neither a sound nor a symbol. I believe early humans communicated much like those of us who find ourselves in a situation where no one speaks our language. In these circumstances, words are useless and efforts to communicate rapidly descend into pointing, gesturing, and making faces. Such signals are not symbols: they are not empty contrivances whose meanings are arbitrarily ascribed and conventionally shared. They are signs; they look like something. Their forms are designed to convey information about the objects or acts they refer to.

A sign transmits information about some object through its resemblance to the object or event it represents. A sign possesses some perceptual feature that suggests the thing it refers to. It shares some physical likeness with an object or process. It could be a hand gesture, a body posture or movement, a facial expression, or any act that provides information that helps to identify what it refers to. A sound can also be used as a sign. However, as a sign a sound is not "empty;" it must echo some specific noise such as the mooing of cows, baaing of sheep, barking of dogs or the crashing noise made by a falling tree.

The first forms of referential communication most likely combined a range of different kinds of signs—with, perhaps, a few conventionally agreed-upon sounds thrown in as well. The cognitive neuroscientist Merlin Donald (1991) argues that speech, our symbolic language capacity, was preceded by a lengthy signing phase. The hand movements and facial expressions we automatically make when we talk—even when we talk on a phone—are traces left over from that era. Even people who have been blind since birth gesticulate and make faces when they talk. There is evidence, he maintains, that these behaviors are rooted in neurological structures that were laid down in the brain previous to speech. He, along with a number of other researchers (Tomasello 2008), hypothesizes that early hominins used a combination of signs and mimicry, a sort of pantomime, to communicate with each other.

The Sign Is a "Tool"

Because we use our anatomy to make signs, it is not immediately obvious that they are tools which are independent of our organism. Expressive signals are clearly part of an animal's biology. When a creature cries out in pain, fear, or even pleasure, it is not crafting its response; it is using a signal that arises spontaneously and that expresses its organism's feeling state. That is very different from what happens when a creature generates a sign. In this latter case, it is intentionally constructing an instrument to inform another creature about something in the external world—something outside itself. Expressive signals don't have to be learned and they automatically resonate with those who receive them. A signal that seeks to represent a creature's thoughts or perceptions about the world around it has to be intentionally crafted for that purpose. The maker of the sign must construct it so that the person who receives it can understand the message. Anyone who has played charades knows that this is no easy matter.

Like a stone axe, a sign is a tool that is designed to accomplish a particular goal.

Signs Are Very Complex

Signs and technology have a lot in common. They are both devices for giving an individual a capacity that is not built into its biology. The ability to create such devices depends on an ability to see things from different perspectives. Kohler's chimp had to see the pole from the perspective of the problem—was it long enough and narrow enough to get through the hole in the back and push the banana to the front of the cage? In other words, a tool must be designed from the perspective of the problem.[1]

Referential communication also involves looking at things from different perspectives. Here the key is being able to look at one's own actions from the perspective of another. The purpose of a sign is to make one's thoughts or intentions understandable to another. This involves an interactive process in which two or more beings must constantly be aware of how each is reading the signals of the other. Here we see how a theory of mind—the awareness of the great apes that the minds of others work the same way theirs do—is a prerequisite for this form of communication.

Signs may be simpler than symbols, but they are still very complex. No sign is able to exactly mirror the object or process one wishes to represent. Signs are analogies. They make use of one kind of material—a hand position, facial expression, body movement, etc., to represent another, very different kind of substance. They are iconic: shorthand versions of some object or event. Because they cannot truly mirror the object they seek to reflect, their construction always involves a choice about which features should be used to best represent it. This involves a negotiation between the members of a community.

The elements used to compose a sign can be very surprising to people outside the circle of users. Certain Native American peoples developed a system of signs that allowed individuals from tribes that spoke different languages to communicate about simple matters. The sign for whites, i.e. Europeans, was a slash that extended across and beyond the forehead. This was meant to indicate a wide-brimmed hat. Only people who came from cultures that did not make or use such hats would have chosen that sign as a means of signifying someone from a culture that differed from theirs in so many other significant ways.

A sign's capacity to model its referent is actually quite limited. Wrapping both thumbs together and flapping one's fingers would seem to be an easily understood sign for bird. From the physical gesture alone, however, it is not possible to know exactly what this sign refers to. Does it denote every bird or one specific kind of bird? Could it stand for a flightless bird—an ostrich? Does it designate only a bird in the act of flying? Perhaps it represents a butterfly. Or is it a metaphor for flight in general or even, perhaps, for running away from something? Cowardice?

While certain signs are more easily read than others, in general, their meanings are not entirely self-evident. They are always somewhat ambiguous. To serve as truly effective means for communicating, both their forms and meanings must be worked out collectively within the context of a community of users. A capacity to use signs, then, depends upon a collaborative social process.

Motivation and Cooperation

The social dimension touched on above brings us to an interesting and rarely addressed aspect of referential communication. Michael Tomasello, co-director of the Max Planck Institute for Evolutionary Anthropology, is currently doing some of the most exciting work on the origins of language. His efforts to test the capacities of chimpanzees to read and use signs have been very enlightening. One of his experiments sheds light on an unexpected dimension of referential communication (Tomasello 2008). Besides the intellectual prerequisites we have discussed, there are also a set of emotional or psychological factors that come into play.

While in general chimps exhibit exceptional skills in interpreting signs, they have certain "blind spots;" there are certain contexts in which they cannot grasp what appear to be simple and straightforward signs. These blind spots occur when the signs contain messages that are not consistent with their normal behavioral expectations.

Tomasello's experiments show that chimps cannot understand a sign that is meant to help them find food: they are unable to read the efforts made by their keepers to direct their attention to containers where food has been placed. Chimps, whose diet consists of fruit, shoots and other vegetable matter don't cooperate in foraging. They do not assist others in acquiring food, and they have no expectations of others assisting them. Indeed, they are very competitive and possessive about food. Tomasello's chimps were thus unable to recognize gestures as signs when they came from an agent who unselfishly desired to help them discover a hidden food source. Unable to relate to the motives behind such efforts, they could not make sense of the gestures.

Signing is not a relationship between an animal and a sign. A sign is a "tool" that one animal fashions in order to communicate information it possesses to another. To be able to read the information contained in the sign, a being must already understand something about the intentions and frame of mind of its maker. Tomasello's research would seem to suggest that signing behavior most likely developed in a context in which cooperation was becoming vital for survival.

This brings to mind the image of a troop of *Homo habilis* moving away from the trees and venturing out onto the plains in search of carcasses and small game. These creatures were entering into a new and more dangerous world: one that their unaided organisms were not designed to cope with. The problems facing these early hominids, the need for collaborative action, created a framework of mutual interests and collective goals that would not only have necessitated the use of signs, but which also would have made them possible.

Honeybees and Vervets?

Before leaving this discussion on the origins of referential communication, I would like to briefly address some of the challenges raised to this framework. As I mentioned earlier, the fact that honeybees have their waggle dance and vervets can generate signals that distinguish between different sources of danger, has served to call into question the notion that referential communication is a specifically human capacity. I have mixed feelings about this.

On the one hand, we must always welcome new research that reveals the amazing capacities of other life forms. Simplistic ideas about what animals can and cannot do are constantly being refuted. But on the other hand, we humans need to understand what we are: the source and nature of the powers that our ascendance has unleashed.

We do, of course, share a lot in common with other animals; in fact, there is no absolute dividing line that distinguishes our biology from theirs. Nevertheless, there is something dramatically different about us. If we are ever to figure out how to manage our out-of-control growth process, we have to come to terms with what we are—not just what we share with other creatures, but also what makes us different. Efforts to constantly obscure or deny that difference serve no larger purpose—neither that of science, humanity, nor of other species.

The bottom line is that neither honeybees nor vervet monkeys are capable of developing a referential system of communication. In this chapter I have sought to show how complex referential communication is. It is not something that can be built into the biology of a creature. It is based on a level of intelligence that enables a being to fashion a tool that gives it a capacity that is not built into its biology.

Honeybees have less than a million neurons in their tiny little brains. Their brains don't even possess cortical tissue. The cortex of a vervet, based on the studies of other monkeys in the same family (guenons), has about 2.5 billion neurons. A chimp has 6 billion, and a human, 23 billion. Honeybees, whose social interactions depend primarily on pheromones, are not serious candidates for a general referential capacity. They do, however, possess an amazing in-born capacity to communicate a specific piece of information about the external world. This capacity, however, is not learned and it is not built upon a general ability to share this kind of information.

Vervets and Diana monkeys, on the other hand, are relatively intelligent primates. They are not, however, on the same level as the great apes. They do not use tools, cannot recognize themselves in a mirror, and do not possess a theory of mind. In other words, they are incapable of seeing themselves from the perspective of another animal or of imagining how another animal might look at things. That rules out the interactive process that is essential for true referential communication.

Recent research on vervets has established that the different calls are not something the animals work out with one another. The capacity to make and respond to these signals appears to be biologically coded in their organisms. If these monkeys possessed a true referential capacity, we would expect to find evidence of signing behavior—efforts to exchange information. After years of reflection and revisiting the monkeys in the wild, the authors of the original study conclude that there is no evidence that monkeys ever communicate with the intent of influencing another animal's mental state: "the lack of a 'theory of mind' in nonhuman primates constitutes one of the fundamental differences between human language and animal" (Cheney and Seyfarth 1997, p. 265). In other words, their calls are fundamentally expressive in nature.

The nail in the coffin of vervet communication, however, would seem to be the fact that recent neurological research has revealed that the vocal cords of monkeys lack "direct neocortical inputs" (Streidter 2005, p. 324). Monkeys, like apes, do not have voluntary control over their vocalizations.

So much for Dr. Doolittle.

Organization

The development of a referential capacity enabled the early hominins to exchange information about the world around them, and most importantly, to share their goals and intentions. This allowed them to coordinate their behaviors in ways no other animals could. It gave them access to what, at that stage in our species' evolution, was the greatest power available to them—organized collective action.

However, communication is not the only requirement of cooperation. The hominins would also have had to develop the social skills and emotional bonds that would enable the emergence of a new kind of community.

Chapter 9

Large Brains, the Sexual Division of Labor, and a New Kind of Community

In this work, I have emphasized the central role that intelligence played in human evolution. It was the intelligence of our great ape ancestors that made certain extraordinary behaviors—culture—possible, and it was these behaviors, in turn, that altered the calculus of natural selection and set our species on a new and unique evolutionary path. In other words, it took a certain level of intelligence to generate culture, and it was culture, in turn, that placed a higher premium on the enhancement of that intelligence.

This scenario begs an important question—do we need culture to account for selection for intelligence? Wouldn't greater intelligence always be to an animal's benefit regardless of the presence of culture? The more intelligent a creature is, the better tuned-in to its surroundings it would be and the greater the range of behavioral options that would be available to it. All things being equal, this is generally the case. When we look at the big picture—the evolution of the animal world seen in broad strokes—there has been a general tendency toward the evolution of more intelligent life forms.

At the same time, however, intelligence is not an unmixed blessing. It is largely a function of brain size, and big brains create a host of difficult problems. To begin with, from a metabolic standpoint brains are the most costly of all biological organs. In adult humans, our brains consume 20 to 25% of the energy we expend when we are at rest. With great apes, that figure is 8 to10%—and for the great majority of other kinds of mammals, 3 to 5%. Humans, then, pay a big metabolic price for our large brains. For natural selection to have evolved such a costly organ, those bigger brains had to have made an important contribution to our survival—one that could compensate for such a high cost.

The problems created by big brains go far beyond their

hefty metabolic price tags. Large brains give rise to a whole set of complicated biological and behavioral difficulties. That's where culture comes in. While intelligence plays an important role in the life of non-human animals, the key to their survival lies in their physical attributes—their speed, strength, and the various anatomical "tools" that nature provides them with. In the case of humans, our physical capacities—the powers we can bring to bear on a problem—depend primarily upon our cultural constructs. It was the access of our ancestors to cultural resources that ensured their survival. And these various cultural devices and techniques, as we have argued, depend upon an ability to perform certain complex mental operations. With humans, then, intelligence is not just an adjunct to our survival behavior—it is the foundation of that behavior. As a result, it was selected for to a greater degree than with any other creature. That's why our brains are off the charts—seven times larger than that of the average mammal.

However, the fact that intelligence was critical to our unique way of life does not mean that our large brains did not pose serious problems for our species. They did. And, indeed, a major part of our evolution consisted not simply in evolving larger brains, but in finding solutions to the problems they brought along with them.

Big Brains and Motherhood

The difficulties associated with big brains go all the way back to the time when they first made their appearance. The mammals were the first animals to develop big brains. In order to be able to grow those brains, they had to overcome a complex set of physiological, behavioral, and nutritional obstacles. The way they addressed those issues gave rise to a package of features that came to form the core of the mammalian biological system.

The fossil data indicates that around 230 million years ago, the reptile branch that gave rise to the mammals (synapsids) began losing out to another branch that was just coming into its own: the dinosaurs (diapsids). To be able to survive the challenge posed by such formidable competitors, the synapsids withdrew into a niche that had been left unoccupied: they became nocturnal. In addition to developing higher metabolisms—warm-bloodedness—to function better in the cooler temperatures of the night, they also evolved a whole new sensory system in order to operate effectively in the dark.

One part of this new sensory system involved the development of night vision. This consisted in replacing the color receptors (cones) in the retina with light-sensitive ones (rods). Until the extinction of the dinosaurs, almost every mammal was color blind. Even today, most mammals have reduced color-vision capacities. There is, however, one exception. It appears that as a result of their climbing abilities, our early primate ancestors (who looked like squirrels) may have been able to function during the daylight hours even during the Age of Reptiles. This may have given the primate lineage a head-start in re-evolving color vision and help to explain why primates have the most highly developed color vision of all the mammals.

Even more dramatic than the changes in vision, however, was the switch to a sensory framework based primarily on smell and sound. While this transformation involved the expansion of older reptilian auditory and olfactory centers in the brain, these had to be integrated and organized in a new way in order to provide a coherent sensory map to replace that previously based on vision. It is in these early ancestral mammals that we see the first evidence of the cerebral cortex. The cortex was a new layer of neurological tissue whose original function appears to have been to connect and weave together the different parts of the newly evolving sensory system.

Because of the development of this new sensory framework, the brains of these first mammals were twice the relative size of their dinosaurian competitors. Within a short period, those brains doubled in size and produced creatures that were operating with brains that were four times as large as those of dinosaurs. The appearance of these oversized brains, the largest in the animal kingdom, did not bring the early mammals any immediate respect. Indeed, their retreat into the night was associated with a dimunition of body size and the adoption of a diet based largely on insects. Seventy million years ago, the ancestors of all living mammals were small, nocturnal insectivores.

Far from being simply a boon, larger brains posed a host of metabolic, nutritional, and behavioral problems. If we could come up with one word to describe how the mammals addressed these different problems it would be motherhood. The development of a nurturing mother was the principal adaptation that enabled mammals to accommodate the large brains made necessary by their newly emerging way of life.

Mammalian motherhood included a number of different features. One was dispensing with the reptilian egg. The egg, which completely separates the biology of the mother from the developing fetus, was replaced by a network of internal blood vessels that would eventually evolve into a placenta. This enabled the mother to nourish and carry the fetus within her until birth, thus providing a steadier temperature, access to more and better nutrition, and a way to remove the developing fetus' waste products. As a result, the fetus could both develop faster and take more time to grow than it could in an egg. This is critical for slow-growing brain tissue.

Then, of course, there are the mammary glands: the feature that gives mammals their name. By suckling her young, the mammalian mother could nourish a helpless infant, giving its brain time to continue its growth after birth. This enabled it to construct an experienced-based behavioral system before it would have to set out on its own.

And lastly, motherhood required the development of a new set of emotional centers in the brain. The emotional bond that exists between the mammalian mother and her offspring—the feature that keeps her from abandoning or eating her offspring, as most reptiles will do—is based on a suite of neurological developments (the limbic system) that provides mammals with emotional capacities lacking in reptiles.

Animals with large, cortically based brains could not have evolved without the developments that took place in pre- and post-natal nurturing. Motherhood lies at the root of mammalian evolution. The emergence of super-sized human brains, however, was more than even motherhood could cope with.

The Limits of Motherhood

In the past few years, there has been a growing interest in the cost of large brains. Increasingly, the focus has turned to the incredible expense that they represent for the mother. Besides physically incapacitating and rendering a pregnant mother more vulnerable, large-brained fetuses create huge nutritional demands. The mother must eat for two—and for her fetus's brain to develop properly, she must eat well.

The larger the brain of the fetus, the longer the pregnancy required and the greater the nutritional demands placed on the mother. This burden continues after birth since she must consume more and richer foods to be able to produce the quality of milk her brainy offspring require. And, the larger the brain of the infant, the longer it must nurse and the longer it takes to mature. The result of all of this is that the larger the brain of the species, the greater the burden on the mother.

In a groundbreaking study based on data drawn from 1247 different mammalian species, Karin Isler and Carel van Schaik, research anthropologists at the University of Zurich, were able to document the effects that large brains have on an animal's reproductive life (Isler and van Schaik 2009B). Larger-brained species have fewer offspring and their births are spaced further apart. There is, then, a trade-off between larger brains and lower birth rates, leading Isler and van Schaik to conclude that at a certain point, in spite of gains in intelligence, larger brains can cease to make a positive contribution to the survival of a species.

This research helps to clarify the issue of why animals do not automatically evolve larger brains and points to an important idea: that the intelligence of an animal is part of a precariously balanced biological/behavioral equilibrium. For an animal to develop a larger brain, it must overcome a set of complicated physiological, nutritional, and behavioral obstacles. An increase in brain size, then, only takes place when the contribution that larger brains make to a species' survival compensates for the problems that accompany them. And most animals, it appears, are smart enough to do what they have to do in order to survive.

Special Problems Posed by the Human Brain

As we get a better understanding of the difficulties created by larger brains, the enormous expansion of the hominin brain becomes that much more remarkable. Within the span of a million years—from *Australopithecus* to *Homo erectus*—the hominid brain, already the largest in the animal world, doubled in size. Besides all the problems that such an unprecedented growth of the brain would have posed under any circumstances, this development was made even more difficult because of our unique form of locomotion. Our species suffers from what paleoanthropologists call the "obstetric dilemma." The evolution of large brains and the development of a hip structure necessary for efficient bipedal locomotion are at odds with one another. Getting a large-brained fetus through a rigid, narrow pelvic opening is a major challenge. A substantial part of our species' evolutionary capital was invested in addressing this issue.

Evolution had two strategies for solving this problem. One was to have the fetus born at an earlier stage of development—while its brain is still small enough to get through the birth canal. This adaptation is clear with contemporary humans; the brains of our newborns are only 25% of adult size. In most mammals, that figure is 80%. The other way to solve this problem was to widen the pelvis of the female—enlarge the birth canal. This is another feature we modern humans possess. And even with both of these in place, childbirth still presents serious risks to the mother.

While these solutions are not perfect, they seem to be the best evolution could come up with. However, they left us with some collateral damage. Women have hips that are not as well designed, biomechanically, as those of men—a limitation that women athletes have shown not to be of great significance. A much more serious problem is posed by the premature state of our infants—the fact that our species must care for the most helpless, dependent and slow maturing newborn in the animal world. To deal with this, we needed to make some further biological—and as we shall see—behavioral adjustments.

In terms of our narrative, we must ask at what point in our evolutionary process did the problems created by big-brained offspring become a significant factor? When did they begin to play an important part in shaping the biology and behavior of our early hominin ancestors?

Given the paucity of the evidence—we have the remains of only one fossil *Homo erectus* infant—it is not possible to directly determine, with any degree of confidence, what the brains of their newborns were like. However, there is a growing body of indirect evidence that gives us a pretty good picture of what was going on. Based on advances in knowledge of how teeth are formed—the way enamel is built up over time—along with evidence drawn from the pattern of tooth eruption and the timing of bone fusion, paleoanthropologists have been able to get a general picture of childhood development in these early hominins.

The evidence suggests that while the brain was much larger, the state of infant brain development at birth in *Homo erectus* was not significantly different from that of their australopith ancestors. Because the size of their brains did not yet require delivery at a radically undeveloped stage, their newborns were not as helpless as ours.

However, their brains were large enough so that the other part of the solution—that involving the widening of the female birth canal—had indeed taken place.

Homo erectus is the first hominin in which we find evidence of a sizeable differentiation between the pelvis of males and females. The discovery of a 1.2-million-year-old pelvis from Gona, Ethiopia, indicates that the female *H. erectus'* pelvis was shaped quite differently from that of the male, "demonstrating the distinctive obstetric anatomy required to deliver a large brained offspring" (Simpson et al., *Science*, November 2008). The pelvis was much broader and deeper than that of the male, which, according to Simpson and colleagues, would have had an effect on the efficiency of their locomotor capacities. While the male *Homo erectus* may have been built for long distance running, the females apparently were not.

Even though *Homo erectus* infants were not born at a radically different stage of development than their great ape ancestors, that does not mean that infant dependency and a prolonged childhood were not major issues for these early hominins. Indeed, these would have posed much greater problems for them than for the great apes—which have the longest childhood dependencies in the animal world. As Carel van Shaik points out in his study of free-living orangutans, a young orangutan is not able to provide for itself until sometime after 8 years of age. It is a huge burden for a species to produce young that are vulnerable for such a lengthy period of time. Van Schaik argues that the great apes, because of their ability to operate in a protected environment high above the forest floor, could afford such a slow maturation process.

Such an extended childhood would have posed a more serious problem for a ground-living being. Even without an increase in physiological dependency, the vulnerability of their offspring to ground-living predators would have posed a serious problem for *Homo erectus*. Although their brain development at birth may not have been significantly different from that of the great apes, there is new evidence that indicates that *Homo erectus* young did, in fact, mature at a slower rate.

A Word about Protracted Childhoods

The reason for extended childhoods has been a subject of much research and debate. It has long been recognized that slow maturation rates and large brains tend to go together, but how and why they are connected has not been clearly understood. Now, as the result of new technologies for imaging and monitoring brain chemistry, we have a much better picture of how these two are connected.

The developmental process of the brain during childhood—not just the generation of more tissue, but the creation of new synapses and the pruning away of unused pathways—uses a huge amount of a child's metabolic resources. While as previously noted, the brain of an adult human at rest consumes 20 to 25% of our energy budget, that figure climbs to 66% in a five-year-old. A research team led by Christopher Kuzawa (*Proceedings of the National Academy of Sciences*, September 2014) has concluded that the human organism does not have sufficient resources to simultaneously build both the body and the brain at a rapid pace—so it cuts back on the growth rate of the body in order to invest more energy into the construction of the brain.

Kuzawa's team compared human growth rates not only with those of great apes, but also, to the degree the evidence permitted, with those of fossil hominins. They concluded that while the stage of brain development of *Homo erectus* might not have been very different from that of the great apes at birth, their larger brains did, indeed, require a longer childhood: "a more extended period between weaning and sexual maturity began to appear at least 1.5 million years ago in *Homo erectus*" (ibid., p. 13013).

The brain of *Homo erectus* was almost a third smaller than ours, and thus, the delivery of an infant did not require the full package of adaptations that would be necessary at a later stage. Nevertheless, the evidence indicates that the large brains of their new borns did, indeed, create problems at birth (as indicated by wider hips), and that their extended childhood dependency would have placed extraordinary demands on their ground-living mothers—and on the community as a whole.

Nutrition

There is another important piece to this large-brain puzzle. As East Africa entered into a drier and cooler phase around 2.9 million years ago, the rich and varied kinds of tropical vegetation previously available to the australopiths became scarcer, and it became necessary for them to make greater use of the foodstuffs that existed at ground level: grass seeds, certain kinds of herbaceous plants, nuts, tubers, corms, bulbs, other root growths—and meat.

During the last few years, there has been a great deal of research into the nutritional implications of the dietary switchover that took place during that period. That was the time—2.9 mya to 2.4 mya—when the brains of the early hominins were increasing by around 35%. While this enlargement of the brain was once simply taken for granted, we are now increasingly aware of the monumental scope of this development and the profound impact it must have had on the nutritional needs of the evolving hominins.

New research reveals that the vegetation that grew on the floor of the savanna was very poor in two essential nutrients: two long-chain polyunsaturated fatty acids that play a critical role in the construction of the mammalian brain (Leonard et al. 2010). On the other hand, the flesh of the herbivores that populated the plains were good sources of both of these vital biochemical constituents. It is difficult to imagine how the increase in brain size could have taken place without the addition of meat—particularly animal fat— to the diet.

It should be noted, however, that making meat a major part of a great ape's diet was not a simple or automatic matter. In the wild, somewhere between 5% and 10% of a chimp's diet comes from animal matter (mainly invertebrates—insects). If that amount increases too steeply, they develop hypercholesterolemia and cardio-vascular disease. There is strong evidence that a genetic change—traced to an allele called E3—was essential for enabling early hominins to consume greater quantities of animal fats without suffering serious health problems.

Fire?

The bio-anthropologist Richard Wrangham (2010) has looked at the problem of diet and large brains from a slightly different angle. He argues that the human digestive system—which lacks the long colon of our herbivorous ancestors—could not have extracted sufficient nutrients from raw foods to support the large brains of *Homo erectus*. If eaten raw, he points out, only 50% of the nutritional value of starchy roots (potatoes, etc.) is able to be absorbed through the human intestine. He argues that this is true, though perhaps to a lesser degree, for meat as well. This leads him to conclude that fire—cooking—was essential for unlocking the nutritional potential of the savanna biota. *Homo erectus*, he argues, could not have evolved a brain that was twice the size of its australopith ancestors without fire.

While this idea has enjoyed a good deal of support, I find it somewhat problematic. The strongest evidence for fire comes from two finds. The most ancient of these, discovered in 2012, is from what looks like a million-year-old fire pit in Wonderwerk Cave in South Africa. The fire pit has a thick layer of ash, and the fact that it is in the mouth of a cave lends weight to the argument that it is not the product of natural causes. The other main piece of evidence comes from an 800,000-year-old occupation site in Israel. The ash contains remnants of bones, giving credence to the idea it was used for cooking. Nevertheless, we don't have what could be called incontrovertible evidence for fire until 500,000 years ago—from a *Homo erectus* site in China.

There are good reasons to be cautious about claims for early fire use. Had fire really been critical to their survival, one would expect, in the words of the eminent paleoanthropologist Ian Tattersall, that "there should be more evidence of it" (Tattersall 2012, p. 112). It is also worth noting that while *Homo erectus* came on the scene around two million years ago, the first evidence of fire (Wonderwerk cave) comes almost a million years after that date—indeed, towards the very end of its run, when its brain had moved up another notch in size. We don't have enough evidence to really know what *Homo erectus* was capable of. However, the simple nature of its stone technology and the fact that it was not able to attach a point on a spear, leads me to doubt whether it possessed a true pyro-technic capability.

The argument for fire is not based on any hard evidence, but rather, on the supposition that without it there would not have been a sufficient nutritional base for the large brains of *Homo erectus*. This assumption raises the question: What about *Homo habilis*? *Homo habilis* also had evolved quite large brains—and it is hard to imagine that those semi-arboreal hominins could have controlled fire.

The main point I take away from all this research is that some important new nutritional ingredient was necessary to enable the early hominins to develop their oversized brains. Whether they had fire or not, that something new was almost certainly meat. We know from the tools left in carcasses and the cuts on bones found in the living sites, that meat was becoming an important part of our ancestor's diet by 2.6 million years ago. It is hard to imagine how our lineage could have reached a hominin level of development unless hunting and scavenging had become an important part of its life process.

Food Sharing

When we put all this together, we can begin to appreciate the difficult problems that were shaping the social and reproductive lives of the evolving hominins. Their larger brains were creating tremendous nutritional and care-taking demands for the mothers. Meat, the food source that had become critical for the evolution of these new brains, would have been almost impossible for those mothers to attain through their own efforts, at least in the latter stages of pregnancy and the early period after childbirth—precisely the times when the rapidly developing child's brain is most in need of high-quality nutrition.

I think it is safe to assume that *Homo erectus* could not have evolved were it not for the emergence of a new kind of relationship between the mother, her offspring and the larger community. Supplying mothers and their dependent children with meat would have been critical to their survival, and, in turn, critical to the long-term survival of the community as well. Indeed, in a different study, (2009A) Isler and van Shaik concluded that the ability of the mammalian mother to provide adequate nutrition to their offspring is a limiting factor in a species' brain size, and that a communal provision of food has enabled certain social carnivores and birds to develop larger brains. This would have been doubly true for our big-brained ancestors.

While we can only speculate about how *Homo erectus* solved these problems, it is likely that certain features we find among later humans began to take shape at this time. The inescapable burden of bearing large-brained, slow-developing offspring would almost certainly have led to a division of labor between males and females: the hunting-gathering division that was more or less universal among modern-type humans before the invention of agriculture.

This division of labor could not have taken place unless food sharing had become an integral part of these early humans' lives. Food sharing—the division of the results of collective effort—would have been necessary to ensure participation in cooperative hunting. At the same time, as a number of authors have pointed out, many of the foodstuffs found at ground level—tubers, seed, nuts, etc.—require preparation even without fire (grinding, pounding, etc.) before they can be consumed. The hunting-gathering division between females and males would have allowed the development of specialized skills and would have greatly elevated the efficiency of food production. But it could only have taken place if the different actors were willing to share the products of independent effort.

Practical considerations alone, however, appears not have been sufficient to create a sense of collective responsibility for childrearing. Added incentives were apparently required. This was most likely the period in which sexual arousal became de-coupled from the estrus cycle and transformed into a continuous and powerful glue between the sexes.

One of the great engines of human drama may well have sparked to life at this early stage in our species' development. A robust and perpetual sex drive almost certainly played a positive role in integrating males into the food-sharing and child-rearing process. However, sexuality is a double-edged sword: a source of competition and passion-filled conflict as well of affection and mutual support. While we have no idea how all of this played out in the case of *Homo erectus*, we know that among *Homo sapiens*, cultural devices—marriage rules and incest prohibitions—have served to regulate sexual access and link it to a whole set of social obligations and expectations.

Cooperation

In this chapter, I have drawn a connection between large brains, tool using, meat eating, food sharing, and a sexual division of labor. All of these are different, interdependent facets of an emerging cooperative social framework.

Co-operation, in fact, is humanity's greatest power. It holds the key to a cultural way of life. But it could not get very far before the evolution of a language capacity.

Chapter 10

Evolutionary Transitions

There were two main phases to human evolution, each of which posed a different, but equally difficult challenge for evolutionary theory. The first, which is what we have covered up until this point, involved the transition from great ape to hominin—from australopith to *Homo erectus*.

Homo erectus was no longer a species of great ape; it was the product of a unique kind of evolutionary process. While some have sought to understand its decent from the trees and its adoption of a predatory life style as simply an adaptation to a changing environment, to life on the savannah, everything about that adaptation depended on the evolution of a set of unique behavioral capacities.

Homo erectus survived by means of cultural behaviors: tool using, communication by means of signs, and an unprecedented ability for social coordination. However, I do not believe that *Homo erectus* should be described as fully human, nor would I characterize its way of life as fully cultural. Evolution is a process by which one kind of creature evolves into something different. There is always a point in this process when it is not possible to determine whether something should be categorized as the kind of being it was, or the kind of being it is becoming.

Homo erectus was clearly not a great ape, but, as I shall argue, it lacked an essential feature of a fully cultural being. It was a halfway creature. The term that best describes *Homo erectus* is hominin. It is definitely a member of our unique, culture-bearing lineage, but human is a term that I believe should be reserved for the very last stage of our species' evolutionary process.

Before the rise of sociobiological thinking, the term that was used to refer to all members of what was seen as a distinctively human lineage was hominid. In those bygone days, it was considered essential to distinguish humans from great apes because humans were assumed to be radically different from other creatures. However, times have changed. As that idea came under increasing criticism, the pendulum swung in the other direction. It has now been decided that humans are really just a kind of great ape and that we should be included in the same family as them. The term hominid (family Hominidae) has come to include all the great apes—living and deceased—as well as every kind of human.

Our species even lost claim to being a sub-family; the term hominin now refers to every member of the great ape lineage that broke off from the chimpanzee tribe some 5 to 7 million years ago and that eventually led to us. It thus includes *Ardipithecus*, *Australopithecus*, *Paranthropus*, and any members of this lineage that have not yet been discovered.

This is bad taxonomy; it tilts the whole debate about humanity's place in nature towards those who would deny our distinctiveness. The need to recognize that we are different has nothing to do with "bragging rights;" it's about understanding what we are—the different forces that shaped us and the challenges we currently face as a result of our distinctiveness. Humans are not just another kind of animal; we won't make it if we don't come to terms with what is different about us. We have to figure out how to control the powerful forces that have been unleashed by our unique cultural way of life. *Homo erectus* was on the path to becoming human, but it had not yet arrived. It depended on cultural skills and on a uniquely organized community for its survival, but its capacity for cultural development was still constrained by certain biological limitations. At that stage in our species' development, culture could not build on itself: advances in cultural capacities still depended on biological changes—primarily on the growth of the brain.

Homo erectus' culture was very primitive. There is no good evidence it knew how to generate and control fire; it wasn't able to haft a stone point on a tool or weapon; it could not work bone, antler, or ivory; it couldn't sew clothes; and there is no sign of art or rituals of any sort. Most importantly, while it could use signs to communicate better than any other animal, it did not possess a true language.

From one perspective, *Homo erectus* could be looked at as a culture-bearing animal; a creature for which cultural skills were simply means to assist it in achieving its survival ends.

At the same time, however, it is necessary to recognize that *Homo erectus* was very different from any animal with which we are familiar. It possessed a brain that was twice the size of that of any great ape and it depended on a social foundation that involved food sharing and a level of communication and cooperation that was a notch above that of any other animal. I suspect that some of its behavior would have been recognizably human.

It appears to have been a relatively successful creature. It migrated from Africa and populated the temperate and tropical regions of the entire Euro-Asiatic landmass. After inventing an Acheulean level of stone technology sometime around 1.5 million years ago, it appears to have achieved some kind of stable equilibrium. There is no indication of any significant cultural advances or evolutionary changes for some 700,000 years.

Then, around 600,000 years ago, a new hominin appeared in Europe. Its roots are assumed to lie elsewhere—almost certainly in Africa. It clearly had evolved from *Homo erectus*, and for the uninitiated, it is hard to distinguish from the latter. But it has a significantly larger brain—1250cc as compared to 1000cc. With this new hominin—known as *Homo heidelbergensis*—we can begin to see evidence of a new evolutionary development. A new evolutionary phase had begun.

Homo erectus was the culmination of a process that saw a great ape evolve into a hominin, that is, into a being whose biology had been shaped by selection for cultural capacities and that depended on cultural behaviors for its survival. *H. heidelbergensis* represented the opening of a new phase; the transformation of a hominin into a human, that is, into a fully-evolved cultural being.

The Path to *Homo sapiens*

We would expect that the dynamics of the second phase of human evolution would be easier to understand than the first. After all, the evidence is much fresher and the way of life that was evolving is more familiar. Indeed, we are looking at the final stages of a process that ended with us. How difficult can that be? Very difficult!

It has been, in fact, much more difficult to make sense of this last phase that the first. The first gave us our initial definition of culture. Our great ape ancestors were able to build a life on the basis of capacities that depended on the properties of things that were not part of their biological inheritance. That is the essence of culture—and it distinguished them and set them on a unique evolutionary path.

However, human evolution was not just a continuous, incremental process. Something dramatically different happened around a 100,000 years ago. While the first phase of human evolution saw the transition from great ape to hominin, the second phase witnessed another transition—from hominin to human. That's a surprising juxtaposition of terms. Aren't we contemporary humans a kind of hominin—members of the same family as all those culture-bearing creatures that evolved out of our australopith ancestors? Indeed, the terms human and hominin are often used interchangeably.

The problem here is that the zoological categories we use to parcel up the animal world are not adequate to handle human realities. The emergence of contemporary-type *Homo sapiens* was not simply another incremental step along the same path our species had been following previously. It represented a radical departure—one that placed humanity beyond the framework of the animal world. Contemporary-type humans do not simply use culture as a means to attain biological ends. We are the products of a fusion of these two different dimensions—a process that transforms our ends as well. Understanding what this transformation was all about is the most difficult problem in human evolution. It is also the most important; it holds the key to understanding what kind of being we have become.

Chapter 11

The Emergence of *Homo sapiens*

Tracing evolutionary connections gets more complicated during the second phase of human development. This is not simply because the developments that were taking place during this period are more difficult to understand, but also because the first phase had changed the evolutionary landscape.

In the first phase, we could focus on Africa. This is where the hominin lineage was born and where it took its critical first steps. However, as these early hominins migrated to Asia and Europe, they gave rise to different populations centers that were separated by thousands of miles and that became different foci of evolutionary development. What makes this picture even more complicated is that these different populations periodically interbred with one another, scrambling genealogical relationships, and complicating the task of determining who was descended from whom. And then, of course, there are gaps in the data. There are a number of important periods during which certain significant developments took place—particularly in Africa—for which we have no fossil evidence.

Nevertheless, in spite of all the problems, there is general agreement on the main outline of how this phase unfolded. And there is an upside to this more recent phase: advances in DNA technology has made it possible to extract intact DNA—in one case, dating from as far back as 400,000 years. This has helped to clarify some of the relationships between the various ancient lineages.

Homo heidelbergensis

Homo erectus, as we noted earlier, appears to have reached some kind of stable equilibrium around 1.5 million years ago, remaining largely unchanged for over half a million years. Then, around 600,000 years ago, a new hominin-type appeared in Europe.

While there is some controversy about its place of origin, the general consensus is that it evolved in Africa sometime between 800,000 and 700,000 years ago, and spread from there to Europe and eventually to Asia.

Homo heidelbergensis closely resembled *Homo erectus* and was clearly its direct descendant. It was a bit less primitive looking; its teeth were somewhat smaller and its face did not project quite as far forward. However, it still had huge brow ridges, a sloping forehead, and lacked a chin. But it differed in one critical way; it had a brain that averaged around 1250 cc (*Homo erectus* was about 1000 cc). Some individuals fell within the lower range of modern humans (1300 cc). That's a quantum step up from *H. erectus* and would seem to imply a significant advance in cognitive capacities. The evidence for that, however, is not readily apparent. While *H. heidelbergensis* did improve upon earlier technologies, there is no evidence of a dramatic change.

Like *H. erectus*, *H. heidelbergensis* made Acheulean-type tools. They did, however, use a more advanced manufacturing technique. They

had learned how to remove flakes with some kind of "soft-hammer." Rather than simply pounding the stone core with another stone, they used wood, bone or antler hammers and punches. This enabled the removal of smaller flakes and provided greater control in shaping the final product. The result was less bulky, sharper edged, and more esthetically pleasing tools. We also find a bit more evidence for the use of fire—an 800,000-year-old fire pit from Israel that contains charred bones, and a probable hearth from Terra Amata, a 400,000-year-old site on the Mediterranean coast of France. Nevertheless, the evidence is still pretty thin.

While there is no real indication of a technological revolution, it does appear that *H. heidelbergensis* had moved up the food chain. The animal bones associated with *H. erectus* sites are invariably those of small and medium-sized game. At Boxgrove, a 500,000-year-old *H. heidelbergensis* site in England, we find the dismembered carcasses of large and powerful animals. While horses seem to have been the prey of choice, the remains of rhinoceroses, bears, giant extinct elk (*Megaloceros*) as well as other large Pleistocene fauna are also present. It would seem that *H. heidelbergensis* had become a top predator. It undoubtedly possessed more advanced cultural capacities than *H. erectus*. While the stone working may not have been revolutionary, the discovery of a cache of well-made spears preserved in a peat bog near Schoeningen in Northern Germany indicates impressive wood-working skills. The quality of these 400,000-year-old spears is quite remarkable. Made from spruce, they were skillfully shaped; they had been sharpened at both ends and were weighted towards the front like a javelin.

No hominin from this era could attach a stone point to the tip of a spear, and before this discovery, it was thought that without stone points, wooden spears would not have had sufficient penetrating power to serve as effective missiles. It was believed they were used as lances—thrusting weapons. The Schoeningen spears, however, were clearly designed to be thrown. They were about seven feet in length and in the hands of a powerful hominin like *H. heidelbergensis*, would have been quite lethal.

However, it would have taken more than heavy wooden throwing spears to kill some of the animals *H. heidelbergensis* was hunting. It is hard to imagine bringing down a rhinoceros without large numbers of hunters cooperating in an organized and planned way. The hunting prowess of these early humans would seem to indicate improved social capacities. Besides the gross increase in the size of their brains, "all of the brain endocasts of this species that have so far been described are said to show expanded Broca's areas" (Tattersall 2012, p. 137). This could be significant. Broca's Area is believed to have played a key role in the switchover from mimicry and signs to the use of sounds for communication purposes. From this, as well as from other evidence we will examine later, it appears that *H. heidelbergensis* may have taken the first steps towards evolving a speech capacity. This could have played a role in facilitating more effective social coordination and might account for the spurt in brain growth.

No Symbols

Perhaps the most significant indication of *H. heidelbergensis'* cultural limitations is the lack of any clear evidence of symbolic behavior. While the term symbolism is given great weight in paleoanthropology, in fact, there is no consensus about its precise meaning. Symbolic can refer to a multitude of different behaviors, and, as with Supreme Court Justice Potter's famous description of obscenity, you know it when you see it, but it's very difficult to define.

The standard definition of a symbol is "something that stands for something else." This definition works very well in the case of language. When we speak, we make use of sounds (words) to represent objects or events we want to refer to. The sounds, then, stand for something else. However, in anthropology, the term is used much more broadly than this. It is applied to any behavior in which objects or acts are endowed with a meaning or value that is not inherent in their physical appearance or properties. For example, with some people meat is not treated simply as something to be consumed for its nutritional value. Before it can be eaten, it must be ritually treated—made kosher or halal—"permissible." Under these circumstances, meat acquires a symbolic dimension.

Symbolism reveals the presence of a new dynamic: the fact that the world is no longer defined entirely by sensory experience—by perceptual processes or biological need. Evidence for the presence of this new dynamic, that is, that objects are being given some special meaning or value, appears very late in human evolution. Clear examples of this behavior are not found until around 100,000 years ago. This is when we find burials, burial goods, and personal adornments—objects such as pierced sea snail shells, animal teeth, and ostrich-shell beads that were apparently designed to be strung on necklaces or, perhaps in some instances, woven into clothing. The use of mineral pigments to tattoo or paint the body (of corpses as well), or maybe to dye apparel, may well be the oldest of these decorative practices.

While symbolic objects or practices do not have any obvious practical significance, they represent an important behavioral threshold. They are an indication of a new cognitive stage: of the fact that humans no longer occupy a world that is bounded by the senses and the appetites, but rather, one that is constructed by the imagination as well.

Certain isolated instances of what appear to be symbolic behaviors can be found during *H. heidelbergensis*' epoch. However, the eminent paleoanthropologist Ian Tattersall concludes "throughout the period of *Homo heidelbergensis*' tenure no hominid produced anything, anywhere, that we can call a symbolic object . . . they neither thought symbolically as we do, nor did they have language" (Tattersall 2012, p. 142-143).

Homo heidelbergensis Begins to Evolve

Like *Homo erectus, Homo heidelbergensis* spread out over the tropical and temperate regions of the Old World and gave rise to different populations that began to evolve independent of one another. Outside of Africa, beginning about 450,000 years ago, *H. heidelbergensis* began to divide into two different branches. Those living in Europe and other western regions evolved into Neanderthals, while those inhabiting central Asia and the Far East evolved into a people known as the Denisovans.

The Denisovans are known almost entirely from their DNA—from a finger bone and two molars from which very good samples of DNA were extracted. That DNA showed that they were related to Neanderthals and that they had interbred both with Neanderthals and with contemporary humans. Their DNA is found in a number of East Asian peoples, accounting for as much as 6% of the genetic make-up of contemporary Papuans.

Neanderthal

Denisovan

The Neanderthals, on the other hand, are the best known of all ancient hominins: more of their fossil remains have been found than of any other pre-sapiens hominin. This is due to the fact that they were primarily European-based, and Europe was the birthplace of evolutionary theory and fossil hunting. New genetic technologies reveal that they. Like the Denisovans, interbred with contemporary humans. Their genes make up between 1.5% and 3% of the genomes of all non-African peoples.

Both Neanderthals and Denisovans were evolutionary dead-ends. They were side branches, evolutionary experiments that are not part of our direct lineage. It appears, in fact, that they were driven to extinction by the influx of *Homo sapiens* from Africa. However, by breeding with *Homo sapiens* and entering into our genomes, they have become an integral part of our biological inheritance. It is now believed that Neanderthal DNA may have contributed to the immune system of non-African peoples, boosting their resistance to certain diseases that were unknown in Africa. There is also evidence that the Denisovans passed on an ability to cope with extreme altitudes to contemporary Tibetans—an adaptation developed by the Denisovans during tens of thousands of years of life on the Central Asian high plateau.

When we turn to Africa, the evolutionary connections are a bit murkier. It appears there were at least two different branches that descended from *H. heidelbergensis* there as well. For most of this period, the fossil evidence is thinner than in Europe and Asia. One of the African descendants from *H. heidelbergensis* is referred to as a "ghost" lineage; while no concrete fossil evidence of it has yet been found, it left its fingerprints in the genome of contemporary African peoples. It makes up as much as 8% of the genotype of the Yoruba people.

However, the other African lineage is the one we are most interested in. Distinctive looking descendants of *H. heidelbergensis* have been unearthed in Morocco at a site dated to 315,000 years ago (Jebel Irhoud). In many ways these hominins resembled Neanderthals—who were living in Europe at more or less the same time. They had large brow ridges, sloping foreheads, lacked a chin, and their skulls are elongated, that is, extend from front to back. However, these features are somewhat muted: the faces are flatter, that is, project less forward and the teeth are smaller. In spite of their primitive features, these people look more like contemporary humans than any other hominin discovered from this period. As a result, they have been classified as early members of our lineage: *Homo sapiens*.

There is a problem with this classification. The genealogy is almost certainly right; these remains belong to members of the hominin branch that gave rise to contemporary humans. It seems that our lineage had developed its characteristic facial features quite early—sometime before 315,000 years ago. However, lumping all the different members of this lineage together as *Homo sapiens*—the earlier ones being separated from the later ones by hundreds of thousands of years—fails to take into account an important fault line—a critical behavioral change that took place during the last 100,000 years of our lineage's evolutionary development.

The Middle Stone Age

Beginning sometime around 300,000 years ago, as the Neanderthals, Denisovans, and our own lineage were evolving into clearly distinguishable hominin-types, we find evidence of the first truly significant cultural changes since the invention of Acheulean technology. This is undoubtedly related to the fact that these new hominins had evolved very large brains—as large, or in some cases, even larger than our own. While we haven't found a skull of a Denisovan, and thus, have not been able to measure the size of their brain cavity, those of the Jebel Irhoud hominins were around 1375cc—more or less the same as contemporary humans. Neanderthals measured even larger—averaging about 1450 cc. By 250,000 years ago, a number of larger brained hominins with more highly developed cognitive capacities made their appearance—and the Old Stone Age gave way to a new, more advanced cultural stage: The Middle Stone Age.

There were three major cultural developments that distinguished this period. The first involved an improved capacity to manage fire. While there are many claims about the early use of fire, the fact is, there is no good evidence of its widespread and habitual use before this period. The presence of large numbers of fire pits and hearths during the Middle Stone Age indicates a new level of mastery (ignition) of this critical energy source.

The second major improvement was the invention of a new, more efficient technique for making sharp-edged stone tools. Instead of laboriously whittling away at a stone core, this new technology was based upon preparing the core in such a way that a single blow could knock off a sharp-edged scraper, blade or point. This "prepared core" technique (Levalloisian) represented a revolution in stone working technique, and while hand axes (Acheulean) continued to be used, they began to assume a secondary role.

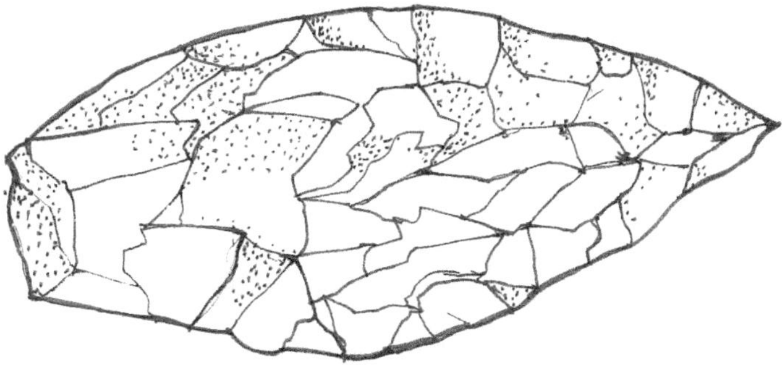

The third important upgrade was "hafting"—the attachment of a stone point or blade to a spear or handle of some sort. This is no simple matter. It requires two things: one, shaping both the tool and the implement to which it is to be connected so that they fit together; and two, figuring out some way to keep them from coming apart—usually the use of an adhesive together with some sort of wrap.

This represented, according to the anthropologist Stanly Ambrose, "an order-of-magnitude increase in technological complexity that may be analogous to the difference between primate vocalizations and human speech" (Ambrose 2001). That may be a bit over the top, but it makes an important point.

The distinguishing feature of the technology from the Middle Stone Age is that it involved greater planning and foresight. It should be noted that these technological advances were not the property of any single human-type. Biologically distinct populations living in widely separated regions were producing them. The Neanderthal and *Homo sapiens* lineages had been separated for at least 200,000 years, yet both had evolved in the same basic direction—larger brains and improved cultural skills. This is an indication that both had been subjected to the same basic selective pressures; both were adapting to a cultural way of life.

Homo sapiens Begin to Distinguish Themselves

Around 100,000 years ago, we begin to see evidence of new and distinctive behaviors associated with members of our *Homo sapiens* lineage—also referred to as anatomically modern humans (AMH). At a site in Israel, we find the oldest burial yet discovered. Along with the skeletal material, the graves possess objects that appear to be burial goods—a boar's jaw, antlers, some stone tools, and what may be the remains of flowers. This is the first evidence we have of ritual: of behavior that is clearly symbolic.

At about the same time, discoveries made at the other end of the continent, in Blombos Cave in South Africa, give added evidence that the behavior of *Homo sapiens* (AMHs) was acquiring a new dimension. From a layer dated to around 100,000 BCE, hundreds of pieces of ochre have been recovered from what looks to be an "ochre factory." Ochre, an iron oxide compound that can be either red or yellow in color, was the principle mineral dye used by early humans. Its function is not fully understood. It is believed to have been used primarily for facial and body decoration and is commonly found sprinkled over corpses in burials. However, it also appears to have been a component of glues used to haft points onto shafts or handles.

The discovery at Blombos Cave is not the first discovery of ochre, but nothing on this scale has been found from such an early period. One of the pieces of ochre has a geometric design incised on it and has been heralded as the earliest piece of artwork yet discovered anywhere. While this claim is hotly disputed, earlier examples of what appear to be artistic expressions are isolated instances that are not repeated and do not appear to become integrated into the workings of the culture. The esthetic experimentation we see at Blombos Cave, on the other hand, would seem to arise from an artistic impulse that became a permanent feature of this new cultural horizon.

However, it is not until around 72,000 BCE that we see the first real expression of our *Homo sapiens* lineage more highly developed cultural capacities. This is when an advanced Middle Stone Age culture known as Still Bay spread across much of southern Africa. In evolutionary terms it was short-lived—lasting about 1000 years. A few thousand years after its disappearance, this region saw another cultural climax called Howieson's Poort.

The stonework of both of these cultural developments was exceptional. The people who created them had learned how to use fire to treat stone to gain greater control over the flaking process. They exhibited a level of skills that had never been seen before: excellent, elongated spear points and a variety of different blades that had been constructed to be hafted onto wooden implements. Of great interest are small "micro-points:" little obsidian flakes that may possibly have been used to line the wooden tips of spears, darts or perhaps, even arrows.

While isolated bone artifacts had appeared at earlier periods, this new cultural climax give us the earliest example of what could be called a bone industry. Bone is very hard to work but it is an excellent material for tool making. It is stronger than wood and can be shaped in ways that are not possible with stone. Not only do we find bone points and scrappers, but also something that that looks like a needle. Needles were essential for making fitted clothing. While fitted clothes would not have been critical for life in Africa, they would acquire greater importance as *Homo sapiens* moved into northern climes.

The evidence of symbolic behavior was unprecedented. Besides powdered and inscribed ochre, we find pierced ostrich-shell and mollusk-shell beads. This cultural efflorescence was the first clear indication that our *Homo sapiens* lineage possessed creative capacities that distinguished them from every other human type.

These two cultural pulses in southern Africa took place around the same time that other closely related *Homo sapiens* relatives began to migrate out of Africa and move into the vast expanses of the Euro-Asiatic landmass. It was not long before they would replace the various hominin types living there. There was something clearly different about us.

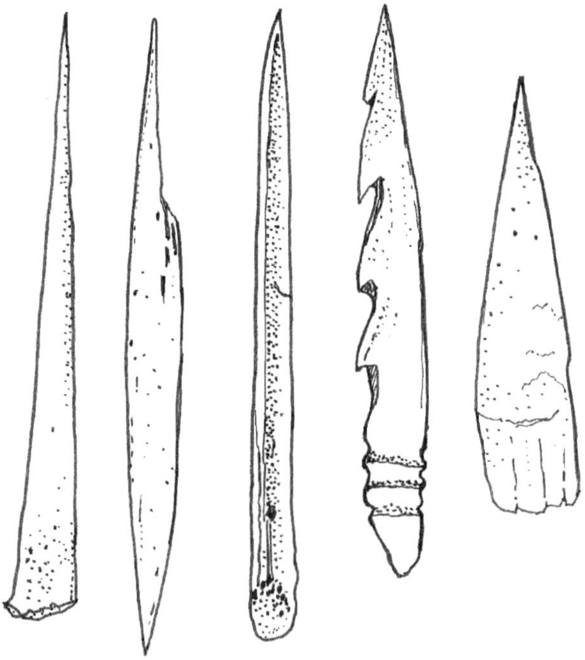

What Was Different?

The traditional view in anthropology was that the dramatic behavioral changes that took place beginning around 100,000 years ago must have been the result of some kind of genetic event that elevated human cognitive capacities. However, with the discovery of early *Homo sapiens* with brains as large as ours—anatomically modern humans—not just at Jebel Irhoud, but at Omo Kibish (195,000-years-

Homo sapiens

old) and Herto (160,000-years-old) in East Africa—the idea took root that perhaps the changes that took place were not due to biological developments, but rather, to cultural processes.

Two major cultural hypotheses were offered to account for these behavioral changes. One was based on demographic factors and the other on the emergence of language.

The demographic hypothesis argues that a critical population threshold is necessary for cultural development (J. Heinrich 2004, 2014). Individuals can only possess so much information and master so many skills. If a community's membership is below a certain critical mass, according to this model, it will not be able to accumulate or—as has happened in some cases—even preserve its reservoir of know-how and skills. On the other hand, when population levels reach a point where there is a redundancy of talent and know-how, it becomes possible to branch out—to develop a division of labor that can support a more elaborate and varied knowledge base. At that point, culture acquires an autocatalytic dimension and begins to build on itself.

The situation in southern Africa during this period lends support to this demographic model. The cultural climaxes described earlier took place during times of exceptionally high precipitation. (Ziegler et al., *Nature Communications*, May 2013). Paleoclimatalogical research has traced these unusual weather conditions to the deepening of the Ice Age in the Northern Hemisphere. As the glaciers moved southward, they pushed the tropical monsoons into lower latitudes, temporarily reversing the long-term drying trend in southeastern Africa. The favorable weather conditions would have produced an unprecedented increase in game and vegetation, providing for a level and concentration of population that could have played an important role in jump-starting the cultural changes we saw during this period.

However, while demographic factors are able to play a role in determining when certain capacities are able to find expression, they cannot account for the presence of those capacities in the first place. Something important had changed in our lineage; the biological constraints that had set certain upper limits to cultural development had been removed. These new *Homo sapiens*—and all those who followed—were able to continuously innovate and improve their cultural outputs without having to make changes in their biology. This was new. Before them, every significant cultural development had been associated with biological changes—a growth in the size of the brain.

The other culture-based explanation traces the cultural capacities that emerged during this period to the invention of language. Language, it is argued, has a transformative effect on human cognition and its appearance could account for the creative cultural advances that take place at this time. Michael Arbib, a scholar with a background in both biology and the computer sciences, has developed the most persuasive version of this thesis (M. Arbib 2012).

The fact that humans can create and master a wide variety of different languages demonstrates that language is not a biological "instinct" but rather a cultural invention—a product of cultural processes. Anatomically modern humans, Arbib argues, must have evolved a biological capacity for language—what he calls a "language ready brain"—before they could have created an actual, effective language system. However, it was the invention of the language system itself that provided humans with the cognitive tools they needed to move to the higher cultural level that we see in Still Bay.

Arbib's ideas actually dovetail with Heinrich's. He believes that the early AMHs possessed a "proto-language"—a mixture of signs and symbols. This system could have worked well in small, isolated, family-based groups where people were in constant, intimate association and did not need a fully developed language system to read each other's intentions.

This proto-language, however, would have selected for a biology that was capable of generating a more highly developed, symbolic-based language system. The realization of this potential would have required the actual invention of such a system. And for that to have happened, something along the lines described by Heinrich may have been required.

It is very likely that the increased population density and the wealth of new social connections generated by the favorable conditions in southern Africa at that time would have brought different family groups and tribes into contact with one another, demanding changes in the idiosyncratic, familial style of communicating. This could have given rise to something like a lingua franka—a simpler, better organized, rule-governed language system.

The problem with Arbib's hypothesis is that the time lag is too large. It is simply not credible that it would take more than 100,000 years for people who were capable of creating a language to actually do so.

Recent discoveries, however, would seem to render both these arguments moot. Exciting new research has revealed that the dramatic behavioral changes that took place during this period depended on biological developments: changes in the brains of *Homo sapiens* that began to take place around 120,000 years ago and were approaching completion about 70,000 years ago. It turns out that the early members of our lineage who had previously been categorized as anatomically modern humans—Jebel Irhoud, Omo Kibish, and a host of others—were not, in fact, anatomically modern.

However, in spite of the fact that Heinrich and Arbib were not aware of this vital piece of biological evidence, I do not believe their ideas were entirely wrong. The scenarios they proposed were quite likely part of the mix. Human evolution involves both cultural and biological dimensions, and the cultural dynamics these authors emphasized almost certainly played a part in the processes that unfolded during the transition from hominin to human.

The New Discoveries

On January 3, 2018, *Science Advances* published a report of new research carried out by a team from the Max Plank Institute for Evolutionary Anthropology in Leipzig, Germany (S. Neubauer et al. 2018). The research established that there was an important biological difference between early *Homo sapiens*—Jebel Irhoud, etc.—and those that evolved after 130,000 years ago. The early types shared a lot in common with those who came later: they had a similar facial configuration and their brains were about the same size. However, adequate weight had not been given to the fact that while their brains may have occupied the same amount of space, those of the earlier members of our lineage were shaped very differently from those that came later.

The study examined the skulls of a wide range of different hominins: *Homo erectus* crania from both Africa and Asia; and fossils of *H. heidelbergensis*, Neanderthals, and a range of early, middle and late *Homo sapiens*.

It turns out that there are two very different variables when it comes to comparing hominin brains. One is gross size; the other is shape. If size is used as the principle criteria, then the brains of every hominin from our lineage could be placed in the same camp—along with those of Neanderthals. However, focusing on shape produces a very different result.

All early hominins, including the large-brained archaics as well as every *Homo sapiens* before 130,000 years ago, had the same basic brain configuration. Their brains were narrow and elongated—projecting backwards towards the rear of the skull. The brains of later *Homo sapiens*—those that had evolved within the last 130,000 years—are very different. Their brains are globular—rounded, not narrow and elongated.

This more rounded brain only began to take shape after 130,000 years ago, becoming increasingly globular until it achieved its full modern condition sometime, according to the authors of the study, before 35,000 years ago. What is particularly noteworthy is that the degree of roundedness went hand in hand with the expression of more advanced cultural capacities.

Analyses of the fossil brain castes have established that the changes that took place in the shape of the skull were the product of changes in the relative size of different parts of the brain. There was a "bulging"—an increase in the size of both the parietal lobe and the cerebellum. This was accompanied by a change in the orientation of the frontal lobe and a shrinking of the occipital lobe—the rear part of the brain. The result was a shorter, more rounded brain. While our knowledge of how the brain works is not sufficient to enable us to understand exactly how these changes affected behavior, nevertheless, by comparing the cultural levels before and after they took place, we are able to get a pretty good picture of what was going on.

Brain Size

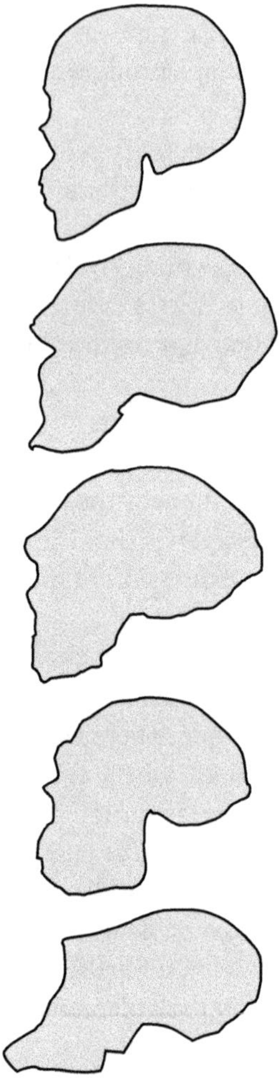

Homo sapiens	1400cc
Neanderthal	1450cc
H. heidelbergensis	1220cc
Homo erectus	1000cc
Australopithicus afarensis	440cc

These changes, I will argue, made possible the emergence of a fully developed symbolic language system. This is not the same as Arbib's proposal. The biological and cultural dimensions of language did not develop at different times; they emerged simultaneously—giving rise to a being in which culture and biology had become indissolubly merged.

To understand what happened, we need to take a deeper look at the nature of language.

Chapter 12

Language: The Emergence of a Symbolic Code

Language is the game changer. It revolutionized the way that information could be exchanged, dramatically enhancing both the social and intellectual capacities of human beings.

Before 100,000 years ago, human intelligence was inseparably linked to biology. Cognitive advances were dependent on changes in the organism, primarily on an increase in brain size. During the biological changes that took place between 130,000 and 35,000 years ago, the connection between biology and human intellectual capacities underwent a dramatic change. The ability of humans to understand the world, to create ever-more complex technological and social forms and to be able to radically change their ways of life no longer depended on evolutionary processes. Humans had come to possess an "open-ended" intelligence: an intelligence that was cumulative, that could assimilate collective experience and enable people to continuously develop their cultural capacities.

While this new kind of intelligence has its own biological foundation, the ability to send a person to the moon was not the product of a biological level of development. It was the product of an accumulation of knowledge (twentieth century science) that was made possible by language. Language transforms human experience into an ever-developing cultural inheritance.

There is no subject that is more complicated than language. In an earlier section, I said nothing was more difficult than intelligence. And that's true. Language, however, is actually another part of the discussion on intelligence: the part that deals with the emergence of a specifically human form of intelligence. What makes this part so difficult is that with language, human intelligence becomes inseparably connected to a kind of cultural tool.

There is a tendency among scholars to see language as either a cultural tool—something separate from our organisms (Everett 2012)—or as the expression of a built-in biological faculty—a language instinct (Pinker 1994). It is both of these, and neither. It is the product of a merger of biology and culture. Because of the immense complexity of this issue, I have divided the discussion into two different parts. For the remainder of this chapter, I will focus on the tool and the evolution of a biology that could make use of it. In the following chapter, I will examine how this tool transforms human intelligence.

Referential Communication

Language is a form of referential communication. It involves an ability to transmit information about things that are separate and apart from the being that is sending the message. This, as we saw earlier (chapter 8), is extremely complex behavior. In order to refer to something outside the self, a being must be able to represent that thing, and represent it in a way that another can understand what is being referred to. This requires a special kind of signal.

Before any members of our hominin lineage possessed a capacity for language, our early ancestors had created a simpler form of referential communication. The key to that system was the use of signs. Signs are representations. They are signals made with hands, facial expressions, body movements, anything that in some way can suggests the object or event one wants to refer to. Animals that lack a capacity to generate signs can only communicate their own internal emotional states: they can only inform others about how they are reacting to events. Our early hominin ancestors, it is believed, used a combination of different signs to create a kind of mime by which they were able to describe objects and events in the external world. This is referential communication, but it's not language.

When we contrast this earlier sign-based form of communication with the system used by contemporary humans, two differences stand out. The first is that the earlier system was based primarily on visual signals—that's what signs are. We, on the other hand, use a system based on auditory signals: sounds—vocalizations. Indeed, the English word for language has the same root as that for tongue: our communication system is generally referred to as speech.

The other notable difference is the way these two different kinds of signals transmit information. A sign is a signal whose meaning is determined by its physical resemblance to the thing it refers to. The signals we contemporary humans use are different. The difference is not simply that they consist of sounds, but that the sounds don't resemble anything. The forms of the sounds have no relationship to their referents. The use of the sound table, for instance, to denote that piece of furniture over there is completely arbitrary. A language could make use of any sound to represent such an object: mesa (Spanish), miiska (Somali), raja (Malay), ban (Vietnamese), ripanga (Maori), or tawila (Arabic).

Signals whose forms have no relationship to their meaning are called symbols. The meaning of a symbol is determined entirely by convention, that is, by the consensus of a community. The use of symbols rather than signs produces a very different kind of communication system.

The Problem of Evidence

Language did not just spring into being from out of the blue, the result of what some, like the linguist Noam Chomsky, believe to have been simply a fortuitous genetic event. Like all major evolutionary developments, it grew out of earlier capacities through a process guided by natural selection. This evolutionary perspective gives us two important pieces of information. The first is that the roots of language must lie in an earlier system of communication—the system of signs that was developed by the first hominins. The second is that switching from visual signs to sound symbols must have conferred some significant advantage or natural selection would not have produced it.

These are both reasonable assumptions. But they raise an important question. How do we know anything about these matters? What evidence do we have that hominins switched from a communication system based on visual signs to one that made use of sound? The biological components of speech, that is, our vocal architecture, consist almost entirely of soft tissue—i.e., non-bony elements. These don't fossilize and thus, don't leave any kind of record. This would seem to rule out the ability to track the development of vocal capacities and thus relegate this whole issue to the sphere of unprovable speculation.

Never underestimate human ingenuity. The soft tissue issue is not the insurmountable obstacle it was once believed to be. It turns out that the vocal apparatus is not the only part of our anatomy involved in the production of speech. Other regions play a role in producing vocalizations, and some of these possess bony structures—and these structures leave markers in the fossil record. As a result, we are beginning to get a picture of when vocalizations entered into the evolutionary picture. Here is a snapshot of what paleoanthropologists are discovering.

During the earliest stages of hominin evolution, a period that embraces the emergence of both *Homo habilis* and *Homo erectus*, there is no evidence that the vocal apparatus of these early hominins had changed in any way from its condition in the great apes. A number of major changes did take place in the skull and face, and the psychologist/neurologist Merlin Donald (1993) argues that the facial bones of *Homo erectus* give evidence of more complex muscle attachments, something we would expect to accompany the use of mime (facial expressions) as a means of communication. However, in terms of vocal anatomy, all the early hominins appear to possess the same sound-making apparatus as the australopiths.

The first evidence of a change in the vocal architecture appears in *Homo heidelbergensis*. Fossil remains from sites dating from 530,000 to 400,000 years ago have brought to light two features that distinguish the vocal apparatus of *H. heidelbergensis* from that of earlier hominins. The first is the shape of the hyoid bone. The *H. heidelbergensis* hyoid is formed like that of modern humans (and Neanderthals), and is different from that of apes and early hominins. This could be significant.

In contemporary humans, the hyoid is a small, free-floating bone that is used to anchor the tongue and to raise and lower the larynx (voice box). It is a part of a complex of changes designed to turn the human mouth and throat into a flexible sound chamber. By enabling the movement of the larynx and assisting in altering the shape of the tongue, the hyoid bone plays an important role in modifying the dimensions of the sound chamber. This greatly increases the number of different tones that can be produced. While chimps, for instance, can make only two distinct vowel sounds, humans can produce as many as twenty. Certain dialects of English make use of as many as fourteen different vowel tones.

While we can't be entirely certain about the role of this

differently shaped hyoid bone, its presence is very suggestive. It would seem to indicate that *H. heidelbergensis* had not only overcome the inability of great apes to voluntarily vocalize, but that the ability to vocalize had become an important part of its survival strategy. The dropping of the larynx, which was critical to increasing the size of the sound chamber, creates a serious danger of choking during eating (think Heimlich maneuver). *H. heidelbergensis* had reached a point where the evolutionary pay off from making a greater number of perceptually different sounds over-rode the drawback of losing a percentage of individuals every generation from choking.

A second feature that distinguishes *H. heidelbergensis* from earlier hominins has to do with hearing. Speech involves more than a special capacity to make sound: it also requires special structures to receive and process them. For reasons that are not fully understood, *H. heidelbergensis* and every hominin descended from it, has a distinctive configuration of inner ear bones (ossicles). The stapes bone is larger than in earlier hominins, a development that is presumed to have something to do with an ability to tune into a specific range of sound frequencies. Humans do appear to have a special ability to hear language sounds. Newborn babies are attracted to speech and people are able to recognize human vocalizations in the midst of noisy and tumultuous events.

There is another skeletal feature that can provide information about the development of a speech capacity. Breath control plays an important role in speaking, and humans have evolved a thoracic musculature whose role is to maintain constant air pressure in the lungs in order to ensure a smooth speech delivery. These breath-control adaptations have placed an increased neural load on that part of the spinal cord that innervates the thoracic region. This shows up as an enlargement of the spinal channel in this region. While the spinal cord itself does not fossilize, the vertebrae that form the canal that houses the cord do. By measuring the size of the canal, it is possible to determine the dimensions of the cord.

The biologists Ann MacLarnon and Gwen Hewitt (2004) decided to investigate whether there might be variations in the size of the vertebral canals among different hominins—something that could indicate the presence or absence of an increased neural load on the spinal cord. The results were consistent with the other findings we have just reviewed.

In the early hominins, the size of the vertebral canal was fundamentally the same as in the great apes. With early *Homo sapiens* (100,000 years old) and Neanderthals, however, the dimensions of the canal were larger—equal to those of contemporary humans. Because of the poor state of the *H. heidelbergensis* fossil material, it was not possible to measure their vertebra. However, because *H. heidelbergensis* was the last common ancestor of both *Homo sapiens* and Neanderthals, and because both these hominin species shared this same anatomical feature, MacLarnon and Hewitt concluded that it most likely had been present in their common ancestor as well. Once again, the evidence points toward some kind of vocal capacity in *H. heidelbergensis*.

Why Sound?

The fact that sound came to replace vision as the main medium of communication implies that sound signals must have an advantage over visual ones. That indeed appears to be the case. To begin with, sound can operate in circumstances where vision is ineffective. This would include nighttime, at long distances, or where vision is blocked or obscured—heavy wooded areas, swimming, etc. One of the most important advantages of sound is that it does not require that those receiving the message have to focus their visual attention on the person sending it. This is related to a point that Darwin himself emphasized: sound allows for "hands free" communication. It makes it possible for people to communicate while they are involved in cooperative activities.

Lastly, it is possible to generate a greater number of perceptually distinguishable sounds than visual signs. The average contemporary language speaker has a working vocabulary of about 20,000 different words, and is able to recognize another 20,000. It would not seem possible to craft 40,000 recognizably different signs. All in all, sound is a better medium for communication than vision.

Why So Late?

The fossil evidence suggests that sound did not play an important role before *H. heidelbergensis*. This raises an interesting question. Why was this? Why did it take so long for sound to become a vital part of human communication? If human evolution began around 3 million years ago, that means that sound was not a major factor for the first 2 million years. Why was it so late in developing?

One issue would have been the need to re-wire the great ape vocal apparatus to enable voluntary control. However, it is difficult to believe that such a development would have taken 2 million years. Indeed, during the first 2 million years of hominin evolution, our ancestors experienced a whole series of far-reaching biological changes. The size of their brains doubled and they were transformed from arboreal vegetarians into culture-bearing, terrestrial hunter/gatherers. It would seem that the changes required to obtain vocal capacities should not have been more challenging than these. But apparently, they were. This raises the question: What was so difficult? What was the barrier that had to be overcome?

When we compare *H. heidelbergensis* with *H. erectus*, two things stand out. The most obvious difference is in the size of their brains—1250 cc compared to 1000 cc. The other consists of the evidence that *H. heidelbergensis* was developing a capacity to use sounds to communicate. I believe that both of these were most likely related.

The principle hurdle to using sound as a tool for referential communication may not have been anatomical, as is generally assumed, but cognitive. Speech is not simply about making sounds. It is about making sounds that convey information about the external world.

Sounds and Symbols

Using sounds to represent objects or events is much more complex than using visual signs. Signs have a relationship to the things they refer to: they capture something about them in their physical make-up. Sounds don't resemble anything but other sounds. They can be used to bring to mind a falling tree (crashing noise) or some animal that can be distinguished by a characteristic sound (lion roaring). But the vast majority of things that humans need to communicate about cannot be represented by a sound: they neither make sounds nor have a sound equivalent.

Sounds have an inherently symbolic dimension. They are not really on the same plane as visual signs. Their meanings are not suggestive or self-evident, but rather, solely the result of decisions made by a community. The fact that their meanings have no relationship to their forms makes them difficult to learn; they must be memorized by rote. When we consider that a contemporary language system requires familiarity with something like 40,000 words, it is clear that a fully developed speech capacity requires the assistance of a biological infrastructure.

Using sound/symbols to represent the complexities of experience raises another problem—one that extends beyond the issue of memory. Because their forms have no inherent relationship to their meanings, it is very difficult to use sound/symbols to communicate a narrative. While it is relatively easy to use individual words to stand for specific things, it is very difficult to string these words together to tell a story.

With mime, the subjects of a story and their activities are acted out; it is obvious who is doing what to whom and what the sequence of events are—these elements are more or less built into the flow of the mime process itself. To be able to represent a set of sounds as a coherent narrative poses a number of difficult problems. It requires a set of rules for organizing and modifying the different sounds to make their status (noun, verb, modifier, etc.) and relationships (subject, object, past, present, etc.) to one another clear. This is what grammar does.

People who only speak one language—as is the case with most Americans—generally don't realize what grammar is and what it does. Most people think of the way we modify and position our words as simply the "right way" to talk—the best way to communicate our thoughts. In fact, grammatical rules, like the meanings of the sounds we use, are largely arbitrary and are determined by convention. In English, for instance, our sentences begin with a subject (Bill), that is followed by a verb (talks), and which ends in an object (Jane). This subject-verb-object pattern is called SVO grammar. The majority of languages in the world don't use this grammatical configuration; they use a SOV pattern. Indeed, the original Saxon language from which English developed had a SOV grammar. We should also point out that different languages make use of almost every possible permutation of these grammatical configurations—VOS, VSO, etc.

Language as a Tool

The principle thesis of this work is that our species is the product of a unique evolutionary process. The difference between us and other animals is that we acquire capacities by making use of the properties of other objects and forces. We did not adapt directly to nature, but rather, to the instruments our species relies upon in dealings with nature. The system of signals that were created to exchange information about the external world was one of those instruments, and the survival of our ancestors largely depended on the social and behavioral capacities those signals made possible.

The first signals were signs—visual representations of the objects and events the early hominins wanted to communicate about. Their ability to exchange this information was critical to their survival, and as a result, their organisms evolved to enable them to become very good at it. The fact we make facial expressions and move our hands when we speak are traces left from that stage of our evolutionary development.

The next stage involved a shift from visual signals (signs) to the use of sound. Just as the human organism had adapted to sign-making, it then adapted to the use of sound as a tool of communication. The lips, tongue, and most especially the throat were re-wired and transformed into incredibly efficient devices for producing a vast range of sounds that could be used to create distinguishable terms.

Because our sound signals do not resemble the objects they refer to, the logic of using sound to communicate leads to the construction of a kind of code. The code makes use of terms (words) whose meanings are arbitrarily ascribed, and as is the case with every code, must be learned by rote. To be able to use these terms to communicate a coherent narrative, the code must include a set of rules for arranging and/or modifying the terms in order to weave them into a coherent narrative (grammar). That's what a language is—a code that enables humans to use a set of external objects (sounds) to stand for inner, inaccessible mental processes—our thoughts.

But what does the evolution of a language code have to do with the appearance of art and ritual—the flowering of the human imagination? How are these related? Can developments in language capacity really explain the dramatic behavioral changes that began to take place around 100,000 years ago?

The Last Step in the Process

At the closing of the Middle Stone Age, the *Homo sapiens* brain went through a reorganization that enabled the mind to assimilate symbols into its inner workings. This made it possible for humans to construct mental models that could go beyond certain limitations of the biological brain. What emerged from this process was a new, culturally organized intelligence—an open-ended intelligence that was able to use symbols (language) to continuously accumulate and improve upon our information base without having to make changes in the biology of human organism.

Chapter 13

Language and Intelligence

What could have happened at the end of the Middle Stone Age? We know that the brain changed shape; we also know that as it changed its shape, human behavior changed along with it. This is when we saw the emergence of contemporary humans—beings with an open-ended intelligence that enabled them to radically change their cultural way of life without having to change their biology. What kind of evolutionary process could have produced such a transformation?

In tackling this problem, we are moving into unchartered waters. The issue we face is not simply about understanding the evolutionary dynamics of some ancient hominin—some distant "other" we call anatomically modern humans or early *Homo sapiens*. This is about us. This is when we became who we are—when our intellectual powers separated us from the rest of the animal world. To address the issue of what happened between 100,000 and 50,000 years ago is to directly confront the question of what it is that makes us uniquely human.

The change that took place during this critical period almost certainly had to do with language. By language, however, I mean something more than is generally understood by that term. The tendency among scholars is to view language as either a cultural or a biological development, and its role in the emergence of contemporary humanity is usually debated within that framework. However, if it is seen as simply one or the other of these alternatives, it is not possible to understand its enormous impact on our species' behavior. Language is neither a biological nor a cultural fact—it is the product of a fusion between these two different dimensions of our being. It is this fusion that gave birth to an intelligence that elevated humanity above the animal world.

There is no scientific consensus on many of the issues that need to be sorted out. However, that does not mean that we are completely on our own: a lot of good research and thinking is being done on this problem. What has been lacking is an evolutionary perspective—a framework that can to pull the different parts of the problem together.

Western Philosophy Starts the Ball Rolling

Western philosophy's model of the mind has played an important part in the evolutionary hypothesis presented in this work. The achievements it made were largely due to the fact that it approached the mind on its own terms. The philosophers accepted consciousness as a reality—that it had form and that it conformed to certain laws. They believed it was legitimate to examine one's own mental processes (as expressed in speech and writing) in order to discover what these laws were. The contemporary cognitive sciences are uncomfortable with such a "subjective" methodology, and as a result have largely abandoned the field of intelligence to those who would reduce it to the operations of a lifeless computing machine.

However, as far as understanding the role of language, Western philosophy must be judged a failure. It viewed language as a direct expression of human intellectual processes, as Hegel put it: "the work of thought." The philosophers saw it as a product rather than as a constituent of a higher form of intelligence. That was an error, and it had consequences for philosophy's development that lie beyond the scope of this work to examine. Nevertheless, in spite of its failure to recognize the part language plays in cognition, philosophy's model of the mind provides a clue that is essential for advancing our understanding of the problem.

Hegel begins his discussion on the workings of the human mind by drawing a distinction between imagery and thought (Hegel 1965). Thought is described as that which is uniquely human: "It is thought which marks the man off from the animals." Hegel does not present thought as simply a gift of the human mind; he sees it as a construct—the product of an internal mental process. Human consciousness, he argues, begins with imagery: we start with a picture of things that is given to us by the senses. This is true, he insists, for animals as well. However, the human mind is not satisfied with imagery: it does not accept the picture that the senses give us as complete. Rather, it re-works those images—processes them— and transforms them into "thoughts."

The question, then, is what is thought? Thought, not just for Hegel, but for Western philosophy in general, is believed to be a way of modeling the world at a level of abstraction above imagery. Our image of a specific cat is not a thought. However, when we use the term cat to refer not just to that individual cat over there, but to all members of the cat family—all house cats, lions, leopards, tigers, etc. –then we have entered into the realm of thought.

The term cat, in this context, refers to a category that includes a number of very different kinds of animals. An image is always an image of a specific thing. There is no image, then, that can represent all the different kinds of animals included within the category cat. This is thought; it operates at a level of abstraction above imagery.

To make sense of this, we need to deepen our understanding of the nature of abstraction.

Abstraction

What is an abstraction? An abstraction, as I use the term, is a simplified model of something.

Western philosophy saw abstraction as the fundamental principle of mind—of intelligence—and not just human intelligence, but that of animals as well. To be able to function in a complex, tumultuous, and ever-changing world, a world in which no two situations, acts, or objects are ever exactly alike, all living creatures need to be able to find commonalities—to discover patterns and regularities that make the world predictable and that enable creatures to develop behavioral strategies.

Living things need to know what kind of object or situation they are dealing with. While we have never seen an animal that is exactly the same shape, size and color as that one over there, nevertheless, its long, thin body and absence of limbs tells us it's a certain kind of animal—a snake.

Kind is an abstraction—a category. It is formed by focusing on a limited number of key elements while ignoring others. Both humans and animals organize experience in this way—into abstract packages. Lions know that all animals that share a certain set of properties are functionally the same—zebras. They also know that zebras, gazelles, and wildebeest are different from one another. By focusing on certain key elements (shape, odor, sound, etc.), they also know that young, old, male and female, large and small wildebeest are all still wildebeests. And what is true for the lion is also true for the wildebeest. They know the difference between lions, leopards and cheetahs—differences that could be matters of life or death for them.

However, while our perceptual categories share a lot in common with those of other animals, at the same time, there is something very different about the way we construct our categories: we don't base them on information that is provided by the senses alone.

A Just So Story

Let us imagine that we have been transported back to an earlier stage of human evolution. A *Homo erectus* hunter is wandering on the veldt and sees an animal off in the distance. The senses provide him with an image that he immediately recognizes to be that of a lion. The hunter returns to his community, and based on the image of a lion he holds in his head, he is able to use mime (signs) to act out the way lions move, the sounds they make, and how they look (mane). This allows him to inform the group of an imminent danger. He uses the image of the animal to create another image that suggests it—a sign. That was as far as his mental processes could take him.

The ability of contemporary humans to parcel up the world is not limited to imagery. While we too would begin with an image of a lion, we are able to go beyond that—to take the conversation further. Recognizing that lions have something important in common with certain other creatures—leopards and cheetahs—we are able to construct another category: feline. Feline includes every different kind of cat. However, there is no such thing as a feline; it does not exist in the real world. It is an abstract model that is constructed out of what is common to a number of different species. There are no individual felines that could provide the senses with materials that could be used to construct an image.

We can take this reasoning further. Felines share something in common with wild dogs and hyenas—they are all predators. This is not a category that corresponds to something we can see and form an image of. It's not defined by the senses—it's a behavioral or functional category. And we can continue in this vein. Mammals include animals as distinct as mice, elephants, whales and bats. There is no image of a real, specific animal that could be used to convey the information that is packaged in the category mammal. And then, of course, there are even more inclusive categories—vertebrate, animal, etc.

Once we got beyond the level of lion, it was not possible to accurately represent these more abstract or general categories with an image. The category lion does, however, possess an abstract dimension; it is built on a limited number of physical features shared by all lions—color, shape, movement, sounds, etc. Lions, however, actually exist; they are a real, biological community—a species. The mind is able to construct a category from sensory information—from certain perceptible features that are shared by all lions. An image of a specific lion, then, because it possesses features that are common to all lions, can serve as a prototype to represent the entire species. Animals are clearly able to construct these kinds of perceptual categories.

Humans are able to create categories that cannot be represented by an image. This is what the philosophers called thought—models of things that are too abstract to be represented by imagery. This was a brilliant discovery. But it posed a problem that philosophy was unable to solve. If thought exists at a level of abstraction that lies beyond imagery—beyond the representational powers of the senses—how are we able to experience it? How can we know what thoughts refer to—what they mean—if those thoughts do not represent anything tangible that exists in the real world?

The Symbol

The ability of the mind to create higher abstractions depends upon its capacity to use symbols.

What is so special about symbols? In many ways they are not all that different from signs. Both share an important feature in common; both are instruments that are intentionally crafted to enable a being to transmit information about objects and events outside itself. Both symbols and signs are constructs created to enable referential communication.

At the same time, however, a symbol is very different from a sign. The feature that distinguishes symbols, that enables them to perform their unique function, is not something they possess; it is something they lack. The secret to the symbol is its emptiness. It is this property, the absence of any perceptual feature that might suggest a relationship to something in the real world, that enables the symbol to transmit a unique kind of information. Symbols provide a way to represent something that cannot be represented by an image.

A sign communicates its meaning through imagery. It bears a physical resemblance to something. Symbols, on the other hand, don't resemble anything: they possess no perceptual information about the things they stand for. This allows them to stand for things whose substance cannot be represented by an image—higher abstractions. Symbols enable the mind to operate on a level of abstraction above imagery; they are the medium of thought.

The issue, however, is not just representation. That would suggest that humans possess abstract models in their minds and are simply using symbols to stand for them and to transmit them to others. The problem is more complex than that. In our little "just so story" about *Homo erectus*, it's not simply that the hunter would have been unable to represent feline or mammal—it's that without a symbol, he could not think feline or mammal. Lacking symbols, he could not be conscious of something that could not be represented by an image. He could only be conscious of the world as it was given to him by the senses. Symbols—words—don't just represent thought; they bring it into existence. This is what Western philosophy missed.

Without language, an animal can only experience the world in sensory terms. Its awareness is limited to how things look, feel, smell, taste, and behave. It can recognize certain general patterns—species and types—based on similarities and contrasts in the physical or behavioral appearances of things. But it can't go beyond that; it can't grasp properties and relationships that are not perceptible to the senses. Humans don't live in a world defined by the senses alone. Our world is a construct, built up of an accumulation of collective mental effort. We live in a city, in a state, in a country, on a continent, on a planet, in a solar system—at a particular time in history. Without language we could not know any of that.

There is, then, a huge difference between the cognitive capacities of a brain that can process language—symbols—and one that can't.

Language is not simply a vehicle we use to communicate our perceptions of the world. Symbols are devices the mind uses to construct more abstract models of the world—to transform perceptions into thoughts.

That is an entirely reasonable hypothesis. The problem lies in proving it—finding empirical evidence in support of it. With humans, language is an inseparable part of all our thought processes. This makes it very difficult to determine what is cause and what is effect; whether language is simply an instrument for transmitting our thoughts, or whether it plays an essential role in constructing them as well. To be able to get a handle on this problem, I have drawn on the experiences of three uniquely gifted individuals.

Thinking Without Words

In 1995, Oliver Sachs wrote a book entitled *Anthropologist on Mars*, in which he introduced Temple Grandin to the world. Temple Grandin is a high functioning autistic woman. She is very intelligent, but was born with serious language deficits. She eventually learned how to speak, but as she puts it, "words are like a second language to me." To understand her mental processes, "you must leave the world of thinking in words . . . I think in pictures: it's been that way forever for me. When I was very young, before any speech/language training, there were no words in my head. Now words narrate the pictures in my imagination, but pictures remain my primary language." (Grandin 2011, p. 107).

As Grandin moved into adulthood, she discovered she had a special connection with animals. She understands how they perceive the world. She was able to broker this understanding into a career designing facilities for holding, moving, and dispatching animals. She can look at situations from their point of view and spot factors that frighten or cause them distress. This has enabled her to create more "animal-friendly" environments.

There are two elements in her writings that are especially valuable. The first is her descriptions of what it means to think without words. She provides us with concrete examples of how effective "thinking with pictures" can be. She is able to plan and organize quite complicated projects in her imagination—in "full-color movies" as she describes it.

At the same time, she cannot process words that represent higher-level abstractions. She cannot grasp anything she cannot translate into an image: "If I have no picture, I have no thought" (Grandin 2006, p. 29).

> Growing up, I learned to convert abstract ideas into pictures as a way to understand them. I visualized concepts such as peace or honesty with symbolic images. I thought of peace as a dove . . . Honesty was represented by an image of placing one's hand on a Bible in court (ibid., p. 17).

Another interesting feature of Grandin's experience is the way she looks at the mental operations of non-autistic people. Sometimes her perspective is surprising—indeed, jolting. She calls those of us who think with words "abstractified," and she points to certain drawbacks that are associated with a reliance on a language-based consciousness. She believes that non-autistic people are unable to perceive certain details with the same acuity she possesses. When normal people look at something, she argues, they see their "idea" of it. Animals and autistic people, on the other hand, "don't see their ideas of things; they see the actual things themselves. We [animals and autistic people] see the details that make up the world, while normal people blur all those details together into their general concept of the world" (ibid., p. 30.

This is a fascinating insight. Grandin is arguing that concepts—the abstract models we construct—shape our perceptions. Animals, on the other hand, see only what is given to them by their senses: in her words, "the actual things themselves." And they see every little detail. Their senses are not constrained by their intellectual expectations. I believe she has put her finger on a real issue: that abstract concepts affect human perceptions. That said, as anyone who has gone into a forest with a trained botanist knows, the knowledge packed into concepts can also heighten one's perceptual acuity.

Grandin, because she possesses a certain level of language, is able to give us unique access to how a mind without language might work. Her mental capacities are, of course, very different than those of non-human animals. She has a huge, *Homo sapiens* brain and she has developed a significant language capacity. While she clearly thinks more like an average human than any animal possibly could, at the same time, I believe she is right in recognizing that her brain functions in a very unique way—as she puts it, "somewhere in between human and animal."

In terms of our discussion, one of Grandin's most valuable insights into her own mental processes is her recognition that images and words represent different levels of abstraction. She is aware that images have an abstract dimension, that is, that an image of a specific object (a dog she liked) can serve as what she calls a "symbol" —what I would call an icon, or prototype, of a general category (all dogs).

However, such icons have severe limitations. An image of a mouse can serve as a representation for all mice, but it cannot effectively represent all rodents (think beavers and capybaras), and most certainly not all mammals or all vertebrates. Symbols—as I use the term—provide a perceptually neutral tool, one that is not connected to a specific image. This enables them to represent categories that are either too abstract to be represented by a specific image or, even more importantly, are not based on perceptual factors in the first place (good, truth, justice, peace, etc.).

The Power of Words

The experience of Helen Keller is another example of how a linguistic handicap can provide insights into the way that language affects the workings of the mind.

When Keller was 19 months old, she contracted a mysterious disease that left her entirely deaf and blind. She had learned a few words before her illness—tea and water (wa-wa)—but she had not learned how to speak. After the illness, she lived in a world of total darkness.

Keller's teacher and care-taker, Anne Sullivan, had been trying to teach her words by having her touch something, and then writing the letters of its name on the palm of her hand. She was not having any success. Keller could copy the finger writing, but she couldn't truly connect words to objects. She was unable to distinguish the word mug from the object that held the milk and the milk itself—indeed, from the act of drinking the milk. Sullivan gave her a new word—water—and decided to take her to the pump-house to see if the experience of feeling the water coming out of the pump, isolated from any container, might make it easier for her to connect the word with the, wet, flowing substance.

In Keller's own words:

> As the cool stream gushed over one hand, she spelled into the other the word for water, first slowly, then rapidly. I stood still, my whole attention fixed upon the motion of her fingers. Suddenly I felt a misty consciousness of something forgotten—a thrill of returning thought, and somehow the mystery of language was revealed to me. I knew then that "w-a-t-e-r" meant the wonderful cool something that was flowing over my hand. That living word wakened my soul, gave it light, hope, joy, set it free (Keller 2004, p. 27).

As Sullivan described the scene: "She dropped the mug and stood as one transfixed. A new light came into her face . . . she dropped on the ground and asked for its name and pointed to the pump and the trellis, and suddenly turning around she asked for my name" (ibid., p. 150).

Something had awakened in Keller: "Everything had a name and each name gave birth to a new thought . . . every object I touched seem to quiver with life . . . I saw everything with this strange, new sight that had come to me" (ibi., p. 28).

Helen Keller's experience provides us with an extraordinary window into the nature of language; one that is made particularly salient when contrasted with that of Temple Grandin. Grandin's senses are entirely normal; her problem is her inability to process higher abstractions—words whose meaning cannot be expressed with an image. She lives in a perceptually organized universe. Ms. Keller's situation is essentially the opposite.

Keller's senses were so impaired that she could not form complete images of things and her world had no clear perceptual structure. When she was able to make a connection between the word water and the "cool, flowing" something, everything changed. The water was no longer experienced simply at a sensory level, but was transformed into something she called an "idea"—a "thought." She immediately realized that everything had a name and that words gave her a path to gaining an organized knowledge of the world.

For the next few years, Keller was obsessed with learning new words, and using these to understand and learn other words. As Sullivan noted, "new words germinate and bring forth new ideas," and new ideas "are the stuff out of which heaven and earth are made."

A Different Kind of Knowledge

What we learn from Temple Grandin and Helen Keller is that language operates on a different level than sensory experience. Grandin's neurological deficits kept her from gaining full access to this level, while in spite of her sensory disabilities, Keller was able to tap into this alternative way of organizing information.

To say that everything has a name, is to recognize that humans encode the whole of experience into symbols (words) and that this enables us to create a system of knowledge that transcends the information provided by the senses. Grandin's deficits pointed to a biological dimension: that without the neurological capacity to process higher abstractions, it is not possible to grasp the meaning of words. Keller, on the other hand, demonstrated the critical role played by the word—an external symbol. It was the word that triggered the dramatic neurological events that took place in her brain. A language capacity, then, consists in the merger of both these domains.

Building Abstract Models

While symbols are necessary for representing abstract mental models, their role does not end there. They not only provide a way to represent such models, but they also play an essential role in the mind's ability to construct them. Western philosophy, as I noted earlier, missed this aspect of the problem. The philosophers saw language as an expression of thought; they failed to recognize it as an essential constituent of thought. Without language our mental models cannot rise to the level of abstraction that is the defining property of thought. It is not difficult to imagine how the mind forms categories based on sensory information; how it can select a core of physical and/or behavioral features that can be used to distinguish different kinds of things from one another—stripes on a tiger, the tawny color or mane of a lion, etc. Sensory data plays a major role in providing the materials out of which we construct such categories as lion, dog, banana, water, etc.

But how can the mind form a model of something that cannot be expressed by an image—something that has no tangible physical form? How do we construct a category like love, justice, or beauty? Or even something like "mammal"—a category that includes such diverse creatures as whales, mice, bats, and elephants? How do we model a narrative—the "decline and fall of the Roman Empire" for instance, or organize an account of why a particular war was fought?

We do this all the time, but we are not aware of what it takes to be able to do this. To begin to get at this, I turn to the work of the Soviet developmental psychologist Lev Vygotsky.

Vygotsky, whose life was tragically cut short by tuberculosis at the young age of 38, did groundbreaking research on the relationship between language and thought. He did not look at this problem from an evolutionary perspective and wasn't asking the same questions I am. He was primarily concerned with educational reform, and to this end he sought to understand how children come to grasp abstract ideas—concepts—and what could be done to enhance the learning process. Directing an institute that carried out experimental research on more than 300 subjects, Vygotsky was able to amass a body of evidence that led him to conclude that abstract thinking—conceptual thought—is not a capacity that can be understood in biological terms alone. It is not simply an expression of some kind of higher, innate intelligence. It is, rather, a faculty that can only develop in association with language. Language provides tools that enable the mind to transform a perception-based consciousness into a more abstract, concept-based one. Vygotsky's great insight was that in learning a language, the child is not simply memorizing a code, but rather, developing its intellectual faculties.

Word Meaning

Vygostky begins his inquiry (1986) by asking the question: What do words mean? What kind of information is conveyed by words? For Helen Keller, it would seem that the word water stood for the "cool, gushing" something that ran over her hand. But she had felt water before—and many other kinds of objects as well. While these had been coupled with words scribbled on her hand, they had not produced the epiphany she had at the pump house. What happened on that occasion that changed her understanding of the meaning of the word water? What opened the door to a new way of experiencing words in general?

While Vygotsky does not appear to have been aware of Helen Keller, his analysis of how words transform mental processes enables us to make sense of her experience. Ms. Keller's understanding of the word water was not altered by a repetitive, mechanical connection between the word and the "cool, gushing" something. Ms. Sullivan had spent endless hours seeking to make a connection between finger writing and various different objects—with no effect. Ms. Keller's epiphany was the result of a new kind of connection, not simply between the word water and the "cool, gushing" something rushing over her hand, but between the word, the "cool, gushing" something, and every other form of water—the hot water used to make tea, the warm water she bathes in, the cold, pure water she drinks, the salt water in the ocean, etc. What the word triggered was a new kind of knowledge—the "thought" of water.

Words, Vygotsky tells us, are not terms we use to stand for the individual things we experience; "a word does not refer to a single object, but to a group or to a class of objects" (Vygotsky 1986, p. 6). Every word is an abstraction; it refers to a category. Indeed, to make words refer to individual things, we have to modify them—the little orange, the orange on the table, the orange with the black mark, etc. The word orange, like the word water, is an abstraction: a general category that includes every different kind of orange.

Verbal Thought

There is nothing particularly original in the recognition that words represent abstract categories. Many others, including Ms. Grandin, have pointed this out. Vygotsky, however, carried this insight a step further than anyone else.

I argued earlier that an abstraction is a simplified model of something. A category like orange, for instance, is constructed by selecting those features that are common to all oranges while omitting those specific features that distinguish individual oranges from one another. That sounds simple enough. The problem arises when we want to create a more abstract category.

Take fruit. This is a category that includes oranges, apples and bananas.

The problem is this. The sensory information— color, shape, taste, texture—enables us to distinguish apples, oranges, and bananas. It tells us where they differ from one another. But how do we represent what they have in common? What allows these different things to be seen as the same kind of thing—to be put in the same category? The fact is, there is no perceptual commonality—no sensory basis for the category fruit.

We might argue that point. Apples, bananas, and oranges can be lumped together as sweet edibles that grow on trees. That is information our senses can give us. But what about lemons and limes? They are not sweet. And strawberries and watermelons? They don't grow on trees. If the term fruit is to encompass what all these different things have in common, it will have to discover that unity outside the framework of sensory experience.

This is where Vygotsky broke new ground. I argued earlier that words (symbols) enable the representation of higher abstractions, i.e., categories like fruit, that cannot be represented by an image. Vygotsky took this idea a step further. The reason we cannot represent a higher abstraction with an image is that images are built out of sensory data—and this is not the case with higher abstractions. That is a startling idea. What other kind of data is there? How can we know what something is unless we can experience it with our senses?

Our senses, of course, are not entirely sidelined. They do play an important part in our capacity to know things. We can't know what the category fruit refers to if we don't know anything about real, concrete examples of different kinds of fruit. At the same time, we never directly experience fruit with our senses. We experience apples, oranges, and bananas—but not fruit. Fruit is not something concrete that can be seen, touched, or eaten. It is an abstract model that is constructed out of what is common to all the different varieties of fruit. Because there is no sensory common denominator shared by all the different kinds of fruit, we need a different kind of model to represent what they have in common: one that is not based on sensory information. What other kind of information is there? What would such a model consist of?

Beyond the Senses

Sensory information is subjective; it is what our senses tell us about the things we experience—how things look, feel, sound and smell. There is, however, a different way of modeling the objects of experience. Instead of trying to define fruit in terms of qualities that are given to us by our senses, we could look at it differently. We could see fruit as something that can be best understood by placing it within the framework of a larger, natural order. Let's try that out.

Fruit are essential elements in the reproductive process of certain kinds of plants. They encase and provide nourishment for their seeds, and they attract animals that consume them and play a critical role in their dispersal. They can only be understood within the framework of a complex set of natural relations.

This is a very different way of describing fruit. This description is not built out of perceptual information; it tells us nothing about shape, taste, color, texture, etc. It describes fruit from a perspective that, according to Vygotsky, does not exist at the level of sensory experience; "it reflects reality in quite another way than sensation and perception reflect it"(op. cit.).

Sensation and thought, according to Vygotsky, are two different ways of knowing things; it takes a dialectical leap to get from one to the other. The ability to make this leap depends upon language—on words. Words enable the mind to make connections that allow it to transcend a perceptual framework and represent reality at a higher level of generality. Without the term water, Helen Keller had no way to represent the unity of all the different, concrete forms in which water is experienced. Without the term fruit, it is not possible to grasp the unity of apples, oranges, and bananas.

By enabling humans to make connections on another level of generality—one that transcend the framework of the senses— language opens the door to a different way of modeling experience. It gives us access to an order in nature that cannot be comprehended at a sensory level. This is the sphere of general laws—of the underlying principles that give rise to a universe governed by cause and effect.

Humans have always known that things fall to the earth, but it would take the word *gravity*--Newton's invention--to be able to elevate that fact to the level of a general principle. Water is not something that is simply "gushing," "hot," "cool," or "clear." Indeed, it is not something that can be fully understood in sensory terms. It has an "objective" dimension: it is composed of molecules that consist of one oxygen and two hydrogen atoms, and it is the properties of these molecules that give water the specific qualities with which our senses are familiar. Without language, the human mind would not be able to model the world at this "objective" level; it would be limited to the picture of things that is given in perception.

Vygotsky himself was not able to fully explore the implications of many of his insights. But he left us with a number of fruitful avenues for research. He saw language as transformational, as providing tools that enable the human mind to construct models of things that go beyond the framework of the senses.

However, it is important to understand that language does not automatically give rise to an objective perspective. Seventy thousand or so years ago, when our *Homo sapiens* ancestors evolved a fully developed language capacity, they knew almost nothing about nature's inner workings or its general laws. They saw nature largely in subjective terms, as a community of intentional agents to whom they were connected through a set of personal relations. But they had evolved a biology and created the cultural tools that would allow them to accumulate an understanding of the "objective" dynamics of the natural world. Little by little, over the course of our historical development, our place in that world would be increasingly revealed to us.

Thought goes beyond sensory information; it seeks to understand things at the broadest levels of generality, within the framework of an overall intelligible order. And thought, Vygotsky argued, is inseparably connected to language: "Thought is not merely expressed in words; it comes in to existence through them" (Vygotsky, p. 218).

The Social and Historical Dimensions of Human Intelligence

The effect that language has on human intelligence should not be seen simply in terms of the mental tools it makes available to individuals. Language is fundamentally a social phenomenon. It doesn't just merge our innate mental processes with symbols, it also merges the minds of different individuals, giving birth to a collective intelligence.

The evolutionary roots of language lie in communication. It grew out of an earlier, sign-based means of exchanging information. It was selected for because it made unprecedented levels of cooperation possible—joint planning, complex coordination, and conflict management. In enabling an efficient combination of human energies, it played a key role in generating the force levels that elevated humans to the top of the food chain.

Recognizing language's genesis in communication highlights its intrinsically social character—and the social character of human thought. Words, as Vygotsky argued, extend the reach of human consciousness. But they are not the property of individuals. They emerge as instruments of communication and their forms and meanings are not private or idiosyncratic, they are the products of an interactive, collective process.

Language gives birth to a knowledge-base that utterly transcends the experience of the individual. It pools the knowledge of people who are separated from each other in both space and time. The greater part of the information we possess is not the product of our own direct experience. Indeed, on what basis do we, as individuals, know that the earth revolves around the sun, that a year is 365 ¼ days long, that our organism consists of trillions of tiny cells or even that Omaha is in Nebraska?

Most of what we know—that the moon causes tides, how internal combustion engines work, that germs cause disease, that air is a composite of different gases, that continents drift, that living things evolve, etc.—is the product of centuries of collective mental effort. Our knowledge base resides in our cultures, handed down from generation to generation and preserved in myths, texts and specialized institutions. As novel as the thoughts of any of us may be, almost everything we know is built upon a foundation of thousands of years of accumulated human experience. It was not some innate human intelligence that put a person on the Moon. That achievement was, rather, the product of an intelligence that was able to assimilate the conceptual framework developed by twentieth century science.

The intellectual achievements of individuals cannot be understood in isolation from a particular level of historical development.

Open-Ended Intelligence

One of the conclusions that follows from the analysis of the way our mental models are constructed, is that a concept is never complete.

There are two dimensions to this. Animals have perceptual categories; schemas that enable them to parse experience into different kinds of things. These schemas organize perception, but are never themselves the objects of perception. Words externalize our categories—bring them to awareness. This enables humans to critically examine their categories and to modify them—to be able to abandon a concept of fire as the release of phlogiston and replace it with an understanding of the nature of oxidation.

Concepts are constantly being modified. They can never entirely encompass the full wealth of their subject matter. This was not the attitude of the ancient Greek philosophers whose genius opened the door to our understanding of these issues. Plato and Aristotle, for instance, did not see our models of things as constructs of the mind; they saw them as discoveries. They saw learning as uncovering the logical design of things—their "truth." This point is not meant to belittle their genius; the framework that the early philosophers used to make sense of the mind was, given the era in which they lived, breathtakingly brilliant. It was, however, only a first step on the path to understanding the workings of the mind.

Aristotle, in particular, generated a wealth of ingenious concepts—but all were either incomplete or simply wrong. And that was to be expected. Concepts do not exist in isolation; they take shape within a system of concepts—as fragments of a community's accumulated body of knowledge. The development of any idea depends upon the level reached by others.

Aristotle, for instance, had no real understanding of what

the heart was. He believed the heart was the seat of the mind—of intelligence and sensations. (He thought the brain was a cooling device.) No one at that stage of human development knew that the blood circulated in the body, and thus no one was in a position to understand the function or structure of the heart. It was only after 1628, when William Harvey published his findings on the circulation of the blood, that it would be possible to develop an adequate concept of the heart (as a pump). Every concept is a work in progress. Each depends upon the general advance of knowledge, on the development of the entire system of concepts that at any moment define the limits of our consciousness.

While language transforms human consciousness into a social-historical process, we should not conclude from this that the role of individual intelligence and creative effort is somehow unimportant. Living human beings still have to do the heavy lifting: advances only take place as the result of their exertions. Their efforts, however, would be useless without the conceptual tools and information base that are their cultural heritage.

Chapter 14

Speculations on the Role of Language in the Changes that Took Place in the Shape of the Brain

There are two things that stand out about the changes that took place in the brains of *Homo sapiens* during the late Middle Stone Age.

The first is that these changes represented a departure from a brain configuration that had deep roots in our evolutionary past: roots that extend all the way back to our great ape ancestors. The latest research on the brains of the various great apes indicate that the basic shape of their brains is fundamentally the same as that of all our early human ancestors—from *Homo habilis* up to and including Neanderthals and early *Homo sapiens* (Periera-Pedro 2017). While the brains of the early hominins were much larger than those of any great ape, the arrangement and proportions of the different cortical regions are fundamentally the same. Changes in that basic design would seem to indicate that something importantly new had taken place in human cognition.

The second significant fact about these changes is that for the first time in human evolution, the appearance of more highly developed cultural capacities was not associated with an enlargement of the brain. Every previous significant cultural development—Oldowan, Acheulean, and that which signaled the onset of Middle Stone Age—was linked to the emergence of larger brains. This last advance, which represented the most far-reaching development in our lineage's intellectual capacities, was not associated with an over-all enlargement of the brain, but rather, with a change in the relative dimensions of certain different parts.

I believe that these two facts, taken together, lend support to a hypothesis that the changes that took place during this period were related to the consolidation of a fully-developed language capacity. This was when the human brain evolved the capacity to assimilate a cultural tool that allowed it to construct a higher form of intelligence.

The brain itself did not need to grow because language provided it with a means for processing and storing information that did not require more neurological space. By enabling the construction of higher-level abstractions—concepts—language made the packaging and organization of information much more efficient. At the same time, language enables knowledge—conceptually processed information—to be "off-loaded;" to become the property of a community. This makes possible an accumulation of knowledge that does not depend on individual storage capacities.

Because the changes that took place in the shape of the skull were produced by the enlargement of two parts of the brain (the parietal lobe and the cerebellum), to be able to confirm this hypothesis we would need to connect the functioning of these two neurological formations to an advanced language capacity. The neurological and cognitive sciences are not yet at a point where they can do that.

Neither the parietal lobe nor the cerebellum has received the attention devoted to certain other parts of the brain (e.g., the frontal region), and their functions are not all that well understood. However, with the use of advanced imaging technologies, new research is beginning to paint a clearer picture of their functions. We now know that both these regions have extensive connections with language centers, and that one of them, the parietal lobe, is involved at the highest levels of human cognition.

The Cerebellum

The human cerebellum is huge, containing half the neurons in our brain. Its size, relative to that of other animals, is every bit as exceptional as is our massive cerebral cortex. It was long believed to be concerned strictly with motor functions—learning and performing complex physical skills. Its large size in humans was believed to be related to the dexterity required for truly effective tool-using and for the coordination of complex vocal movements. However, new research has revealed that its connections with language centers are more extensive than previously thought.

A recent study suggests that the cerebellum's "regulation of language functions is analogous to its role in motor performance" (Starowicz-Filip et al. 2017). In terms of motor behavior, the cerebellum's role is to predict the kinds of coordinated movements that are necessary to accomplish some task. When we reach for a glass of water, for instance, we don't have to think about how our fingers are positioned, what the wrist is doing, or what's happening with the elbow and the shoulder. The cerebellum plays a key role in synchronizing the movements of these different body parts so that they can work together to accomplish a goal. It's now believed that the cerebellum may perform something similar when we speak.

Under most circumstances, when we talk, we don't have to think about the mechanics of producing a message. We think about the point we want to make and the organization of the words and the synchronization of the different movements of the vocal apparatus take care of themselves. It is assumed that the cerebellum plays a key role in this "automatization" of language behavior.

There are two different issues involved in automatization. One has to do with phonetics—the coordination of all the different parts of our sound-making apparatus. The other has to do grammar. Grammar is much more critical to a fully developed language capacity than is generally understood. It enables a body of independent words—distinct concepts—to be transformed into a coherent narrative. By modifying their forms and determining the position of the different words in relation to one another, grammar enables them to function as elements of a single, integral message. Anyone who has learned a second language knows how mastering the grammar is the key to being able to "think" in that language—not having to consciously translate from one to the other.

While it is not yet possible to establish the role of the cerebellum in the emergence of a grammatical capacity, it should be mentioned that people with certain kinds of cerebellular damage are unable to recognize grammatical errors or use grammatical rules to organize their discourse (op. cit.). Frankly, it is difficult to image what other behavioral change could have been associated with the growth in the cerebellum that took place during this critical period.

The Parietal Lobe

It appears that the bulging of the parietal lobe was not due to the enlargement of the parietal lobe as a whole. Rather, it was almost certainly produced by an expansion of a cortical body buried within it—the precuneus. Because its location makes it difficult to observe, the importance of this cortical body had, until recently, been largely overlooked. We now know that it is a neurological powerhouse. It has extensive connections with a broad range of brain regions, particularly the frontal and pre-frontal areas. It is an important part of the default mode the brain assumes when it lapses into a resting state. Indeed, it is the brain's "hot spot"—the most metabolically active part of the resting brain.

While an understanding of the role of the precuneus is still at an early stage, what is being learned about it is consistent with the idea that it played a critical role in the emergence of a fully developed language capacity. It possesses three features, in particular, that would seem to indicate that.

The first is that the precuneus is a part of the "association" areas of the brain. It has no direct relationship with any of the primary sensory areas. Its role consists in "elaborating highly integrated and associative information rather than processing external stimuli" (Cavanna 2006). In other words, it does not work with raw data, but rather, reworks the products of previous mental operations. The association areas play a major role in processing abstract words. The new imaging technologies show that the precuneus, along with the frontal region, become active when an individual thinks about such words as law, justice or truth—words that cannot be represented by an image. These are precisely the kind of terms that Temple Grandin could not process and which she sought to represent with concrete images (truth by a Bible, etc.).

The second important feature is that even though the precuneus is not directly connected to the sensory system, it is importantly involved with mental imagery. The images it generates, however, appear to be primarily those that take a third person perspective; they present a view of things as if they were being seen through the eyes of another. This enables us to imagine how we appear to others—to see ourselves within a larger framework—and has earned the precuneus the label of the "mind's eye." As a corollary of this, the precuneus plays an important role in "theory of mind" processes—imagining what other people are thinking and how they might look at things. The ability to read the intentions and states of mind of others—knowing what their intentions are— plays a critical part in language comprehension (Tomasello 2014).

The third important feature—a conclusion reached by all researchers— is that the precuneus is involved "in the implementation of a wide range of higher-order cognitive functions" (Cavanna 2006). This is consistent with the idea that it plays a central role in processing higher level abstractions.

Going Beyond Imagery

The word that seems best to describe the changes that took place with the emergence of contemporary-type humans is imagination. By three hundred thousand years ago, that is, at the on-set of the Middle Stone Age, the human brain appears to have more or less reached its maximum size. At that time, we see evidence of advances in intelligence—better stone tools, hafting, and probably fire; but culture is still primarily conservative rather than creative. It's not until after 100,000 years ago that we see the flowering of creativity—not just better made and more effective technologies, but burials and burial goods (ritual), personal adornments, dyes, art and music (bone flutes). A new kind of energy entered into the behavior of our ancestors.

All preliterate peoples known to anthropology have believed that there is another world that exists apart from the one that is given to us by our senses— the one we occupy. This other world is believed to largely account for the order in this one—the daily cycle of the sun, the phases of the moon, the changing of the seasons, the coming of the rain— indeed, the gift of life and the fates of human beings. And every people have sought ways to communicate with that world and to better understand and influence it. This is where the roots of myth and ritual lie.

The belief that there is a world beyond the reach of senses was not simply some primitive delusion. The fact is, there are forces that importantly shape our world and that cannot be grasped at the level of the senses. The error lay not in imagining the existence of such forces, but in thinking that they were outside of nature— and that they were driven by personal, intentional goals rather than by physical dynamics. The appearance of imaginative behavior during the period when the human brain was evolving a new shape, argues that the awakening to a sense of hidden mystery went hand in hand with the emergence of a fully developed language capacity. However, at this stage in our understanding, it is not possible to clearly determine what is cause and what is effect in this process. Did a new level of imagination elevate our linguistic capacities to a new plateau—or did new language capacities stimulate new imaginative horizons? As with most of these either-or kinds of problems, most likely both contributed to the process. While I believe Vygotsky made an important breakthrough by recognizing that our special human intelligence is the result of the fusion of language and the workings of the biological brain, nevertheless, there is still much that we don't understand about how the mind creates and operates with higher-level abstractions.

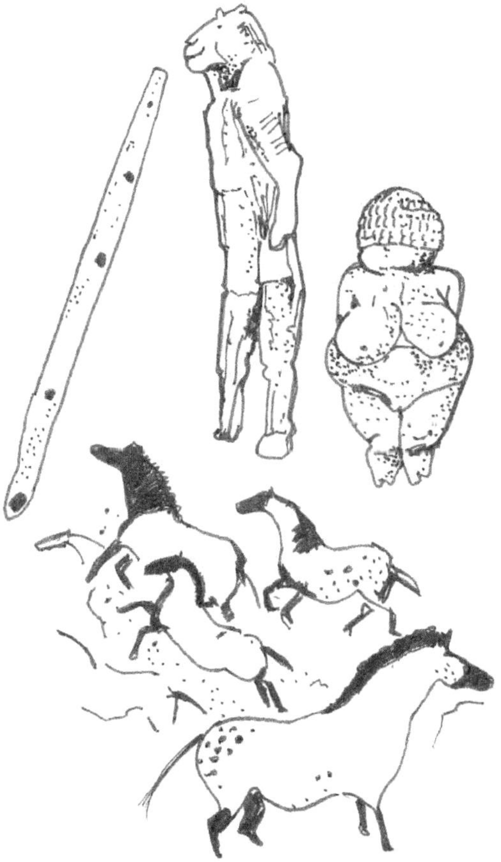

The late Rudolf Arnheim, a brilliant philosopher and art historian (1969), argued persuasively that the mind cannot grasp an idea unless that idea is accompanied by some kind of image. Intuitively, this resonates. He is not arguing that we can form an image that can fully represent such abstract notions like beauty, truth or justice, but rather, that we need some kind of image—something that can serve as a sign or metaphor that enables the mind to attach the concept to something concrete. In the case of beauty, perhaps, an image of something generally acknowledged as beautiful (the

Mona Lisa). A word that has no connection to anything in experience, he maintains, can have no meaning.

At this point, I would like to take a brief look at some of the new work being done in neuropsychology. Equipped with new imaging technologies, researchers have begun to make progress in understanding the very issues we have been discussing—how the brain processes higher abstractions.

The Brain Processes Abstract and Concrete Words Differently

An important article appeared in the March 2005 issue of the magazine *Brain*. The authors, S.J. Crutch and E.K. Warrington announced that working with new technologies for tracking neurological processes, they had attained "the first evidence that abstract and concrete word meanings are based in representational systems that have qualitatively different properties." This discovery initiated a torrent of important research which was summarized by Maria Montefinese in the February 2019 edition of the *Journal of Neurophysiology*: "Evidence from both behavioral and neuropsychological studies suggest that different types of organizational principles govern semantic representations of abstract and concrete concepts." Put in simple language, the brain processes abstract and concrete words differently.

Words that refer to things that can be represented by images, that is, that can be described in perceptual terms, are processed in areas with extensive connections to the sensory system. This would include such words as lion, rock, river, mountain, shoe, etc. Words that are more abstract, that refer to things that cannot be represented by an image—law, justice, truth, etc.—are processed in the association and language areas: "These results indicate a greater engagement of the verbal system for the abstract words and a greater engagement of the perceptual and mental image generation systems for the concrete concepts." (op. cit.)

What struck me about this statement is the idea that abstract ideas are processed in the "verbal system"—in language areas. If we think about it, we can see that this makes a lot of sense. Humans create words to label everything they can distinguish in experience. And each of these words is, at some level, a concept. But some of those concepts are more abstract than others. If I want to know what a rose is, I can be shown a picture of a rose and that would go a long way towards satisfying my curiosity.

If I was to ask what a flower is, showing me pictures of different kinds of flowers would help but would not truly inform me about what they all have in common—what it is that, in spite of all their differences, makes them all flowers. Flower cannot be defined at a sensory level. It is a reproductive organ of certain kinds of plants. That answer involves a functional principle that can only be represented at the level of verbal thought. (Appendix 5)

Take an even more abstract example. If I was to ask what is the Krebs cycle, there is nothing concrete—no sensory data—that would enable me to understand what the word refers to. Google defines the Krebs cycle as "a series of chemical reactions that occur in most aerobic organisms and are part of the process of aerobic cell metabolism, by which glucose and other molecules are broken down in the presence of oxygen into carbon dioxide and water to release chemical energy in the form of ATP."

The Krebs cycle is a concept. It is a model of a chemical process. The model itself is constructed out of other concepts. Every term in the forgoing description of the process is a concept—a model built at a level of abstraction above perception: aerobic, organism, metabolism, glucose, etc. There is no way to communicate, indeed, even to understand the nature of the Krebs cycle except by means of these other concepts. Every concept is constructed out of other concepts, and since concepts cannot exist without words, our models of things are built out of words. This is the real meaning of verbal thought.

Language, then, is the means by which humans are able to model the world at a level of abstraction above sensory experience. This enables us to discover general principles—cause and effect relationships—that are not accessible at a sensory level. The development of language and higher intellectual faculties went hand in hand. The ability to synthesize higher abstractions and the ability to organize discourse into hierarchical grammatical structures are most likely inseparably connected.

A Hypothesis on the Evolution of a Fully Developed Language Capacity and the Emergence of Contemporary Humanity

The fossil evidence suggests that there were two major stages in the development of human language. The first consisted in a signing phase—the use of signs and mime to transmit information about the external world. This capacity grew out of great ape intelligence—in particular, an ability to read the intentions of others and to be able to see things from their perspective (theory of mind).

This stage may have lasted as long as two million years. There is no evidence that the vocal apparatus of the early hominins—*Homo habilis* and *Homo erectus*—was in any way different from that of their great ape ancestors. *Homo erectus*' hyoid bone was identical to that of the australopiths, and its breathing apparatus appears not to have undergone any change from theirs. Sound, it would seem, did not play a significant role in their communication process.

The second stage most likely began with *Homo heidelbergensis*. This is when we first find evidence that vocalizations were playing an important part in human communications. The evidence is thin—a modern-type hyoid bone, a larger Broca's area, changes in the inner ear, and a growth spurt in the brain. However, given their social accomplishments (large game hunting) and the fact that the different hominin lineages that descended from *H. heidelbergensis* all appear to have possessed some kind of speaking capacity, it would seem likely that they were evolving such a capacity.

It now appears that all of the so-called "archaic" humans—Neanderthals, Denisovans, and early *Homo sapiens*—had evolved some form of speech. While we don't have a Denisovan skeleton, their DNA shows they are closely related to Neanderthals, and interbred with both them and contemporary *Homo sapiens*. Neanderthals and early *Homo sapiens* clearly possessed the basic anatomical requirements for some kind of speech: modern-type hyoid bones, breathing controls, inner ear adaptation, and large brains.

These anatomical features were not accidental; they were the products of natural selection. The fact that these hominins had evolved a vocal anatomy shaped by speech-making imperatives indicates that communicating with sound was an important element in their biological success. At the same time, their cultural deficits argue that their level of speech capacity had not yet evolved into a fully developed language capacity. They most likely possessed something along the lines of the protolanguage proposed by Arbib: a mixture of symbols, signs, and mime—similar to how Temple Grandin described her own speech capacities. As she puts it, she thinks in "pictures," and that while "words narrate the pictures in my imagination . . . pictures remain my primary language."

While the archaic hominins may have operated primarily with imagery, I believe that words played a crucial role in their communication process. Sound, as I argued earlier, is a more efficient way to encode information than visual signs. However, the words they used are probably best described as names—labels linked to images rather than to concepts. The huge size of their brains leads me to believe they may have had extensive vocabularies. Their limited cultural capacities, on the other hand, suggest that their brains had not yet evolved the full complement of language skills. I believe they lacked the two essential components discussed above: an ability to process higher abstractions and a capacity to organize their discourse grammatically.

Both of these are critical for a transition from an image-based consciousness to one based on verbal thought—on an ability to think with words. For the archaic hominins, words were tools used to label images; for contemporary-type *Homo sapiens*, words are tools for transforming images into concepts—for creating more generalized models of reality.

The Emergence of a Language-based Consciousness

While the appearance of a fully developed language capacity opened the door to an objective understanding of the natural world, such an understanding does not develop automatically. Concepts are constructs—they do not emerge spontaneously from sensory experience. They must be built—they are the products of a collective, accumulative process of reflection and mental effort.

While our early *Homo sapiens* ancestors possessed the same biological "hardware" we do—the same capacity to construct concepts—they began with very inadequate "software." They were brilliant at making use of nature's resources for practical ends, but they knew almost nothing about its deeper well-springs, and in particular, nothing about who and what we humans are—how we got here and how we fit into the world around us.

Little by little, humanity accumulated knowledge—constructed more adequate concepts. Agriculture, sedentary life, and most importantly writing, would accelerate the learning process. It's no accident that the Ancient Greeks, who were the first people to create a complete alphabetic system, would also have been the first to discover logic—to imagine that nature was governed by universal laws of cause and effect. (See Appendix 4)

Slowly, humanity began to develop models—concepts—that enabled us to better understand how nature works and to begin to recognize that it is bigger than us and that it was not put here simply for our purposes. But even after Copernicus put the Sun into the center of the Solar system, it has been difficult for humanity to abandon its self-centered perspective and to see nature as more than simply a means to our ends.

We have learned more in the last 200 years than in the previous 70,000 years of our species' existence. We have learned about the Big Bang and the expansion of the universe, about subatomic particles and quantum physics, about gravity and relativity, about evolution and continental drift, electricity and digital information, etc., etc. Indeed, our knowledge has reached a point where it is now delivering an unwelcome message—that our growth process is undermining the natural systems that are necessary to sustain life on this planet.

The problem we face is that while our knowledge of how nature works might play an important role in science and technology, it has not truly filtered down and become a part of how the great majority of humans think about their lives. What's missing is a more adequate concept of humanity—one that can connect us to the natural world and that can guide us in reconciling with that world.

Chapter 15

Selfishness, Cooperation and Altruism: E.O. Wilson Changes His Mind

In 2012, the issue of cooperation moved to center stage in the debate about human evolution. This represented a dramatic change. Before then, the issue of cooperation had been raised by a few authors (see Bowles and Gintis 2011), but had not received much attention from mainstream evolutionary anthropology. The oddest thing about this turn of events is that E.O. Wilson is largely responsible for both circumstances: the original attitude that largely ignored cooperation, as well as that which later elevated it to a central place in human evolution.

Competition as The Driving Force

In *Sociobiology* (1975), Wilson put forth the view that competition, not cooperation, was the primary driver of human evolution. While recognizing that humans are social animals and that we receive certain benefits from living in communities, he argued that we are not good at cooperation. Evolution did not shape us to be good team players. We are, in fact, constitutionally selfish: incapable of putting the interests of the group over that of the individual. This self-centered behavior, we were told, is the result of certain inescapable evolutionary dynamics—the inevitable outcome of our vertebrate reproductive process.

The social insects—ants, bees and wasps—are not like us; they are good team players. The worker bees and soldier ants readily sacrifice themselves for the benefit of the hive. This is because, as Wilson explained, they are the products of a different kind of reproductive process than humans. The workers and soldiers of these insect species are members of sterile castes. They are the products of parthenogenesis—a peculiar form of asexual reproduction carried to great heights among certain social insects. The vast bulk of hive members develop from millions of unfertilized eggs produced by the queen. They do not mate and leave no offspring. Whether they live or die has no consequence for the genetic make-up of the hive. Only the Queen's survival has evolutionary significance.

Humans are different. We are vertebrates and vertebrates reproduce sexually. This results in a very different evolutionary calculus. Every member of a vertebrate social group contributes to the genetic make-up of the group. Those that leave the most offspring will make the largest contribution to the group's gene pool and will eventually push the others off the evolutionary table. And who will leave the most descendants? Those, according to Wilson, who do everything they can to leave more of their genes than their competitors. Vertebrate social interactions, whether conscious or unconscious, overtly or covertly, are dominated by the logic of furthering one's own genetic interests: "Selfishness rules the relationship between members" (Wilson 1975, p. 380).

This is a very inadequate view of natural selection. Darwin viewed natural selection as the mechanism by which animals adapt to their environment. Wilson did not present a hypothesis about what kind of environment or way of life our ancestors were adapting to. Instead, he developed a hypothesis about human social behavior that was based solely on this narrowly conceived, competitive, model of vertebrate reproduction. Whether an animal leaves more offspring or not does not depend on the level of its selfishness, but rather, on whether its behavior enables it to better deal with the challenges of its environment. If selfishness weakens the social bonds an animal depends on, or gets it ostracized from its community, selfishness would result in less, rather than more offspring.

Wilson Has A Change of Heart

Then in 2012, Wilson came out with a new book: *The Social Conquest of The Earth*. It represented a major shift in his thinking. He now argues that cooperation rather than competition was the key factor that shaped human evolution. We are not just "vertebrate" in our social behavior, but in many ways, more like the social insects. In a complete reversal from his previous position, he now holds that it is our exceptional capacity for teamwork—our ability to put the interests of the group over that of the individual—that accounts for our species' unparalleled position of dominance within the animal world.

This new view is based upon widening the aperture of his evolutionary lens. Rather than focusing just on the reproductive dimension, Wilson looked at the Big Picture—the way of life our species was evolving. This is what determines what behaviors would have contributed to or detracted from our ancestor's evolutionary success.

The survival of the human individual, he now maintains, depended not on the outcome of individual competition, but rather, on the success of the group to which the individual belonged: "The precursors of *Homo sapiens*… formed well-organized groups that competed with one another for territory and scarce resources…the outcome of between-group competition is determined largely by the details of social behavior within each group … how well a group performs depends on how well its members work together (Wilson 2012, p. 53-54)."

Wilson sums up his thinking with what he calls the "iron rule" of genetic social evolution: "Selfish individuals beat altruistic individuals, while groups of altruists beat groups of selfish individuals" (ibid., p. 243). He no longer sees concepts of honor, morality, generosity, conscience, and the "greater good" as simply the self-serving subterfuges of a calculating, selfish being. They reveal, instead, urges that are rooted in the biology of a creature that has an over-riding survival interest in preserving the solidarity of its community.

This new outlook represented an important step forward in Wilson's thinking. However, it raises an obvious problem: humans are not just selfless worker bees—human behavior is not just about self-sacrifice. Wilson recognizes this, acknowledging that selfishness is also deeply rooted in the human condition. This, he argues, derives from fact that humans, in spite of their social interdependence, are still vertebrates, and as such, cannot avoid competition and selfishness. He ends up with the view that humans possess a genetically fixed, internally divided motivational system. We are torn between conflicting altruistic and selfish instincts.

I have introduced this brief encapsulation of Wilson's thinking into our discussion for two reasons. The first is, in part, to be fair to

Wilson. He was the intellectual godfather of sociobiology and he played a key role in establishing a foothold for biological/evolutionary thinking within the social sciences—and in the public sphere as well. However, the original ideas he formulated back in 1975 were, frankly, crude and poorly thought out. Their impact was largely due to the fact that anthropology and the other social sciences had failed to incorporate humanity's biological dimension into their thinking, and were completely unprepared to respond to this different perspective.

With the *Social Conquest*, Wilson puts forth a model of human evolution that comes much closer to meeting the requirements of an evolutionary hypothesis. For the first time, he discusses what humans were adapting to—what kind of natural selection was shaping our species. As a result, he comes up with a model that is closer to the truth—and that resonates with human experience.

The second reason I decided to include Wilson's new evolutionary hypothesis is for heuristic purposes. While this new model is a step forward, it is still seriously flawed; it based on the same erroneous premise as the ideas he presented in *Sociobiology*. Wilson does not recognize anything unique about humanity's evolutionary process: "Humanity originated as a biological species in a biological world … no more and no less than did the social insects." (Wilson 2012).

There is no mention of culture—of language or reflective intelligence. He looks at humanity from an entirely biological perspective—as the product of the same evolutionary logic as every other creature. In this chapter, I would like to take a different look at the issues of altruism and cooperation—one that recognizes that these behaviors, like everything that humans do, emerge from a fusion between biology and culture.

The Biology of Altruism

We should begin by acknowledging that forms of altruism do exist in the animal world—and that these are not limited only to the social insects. While I will argue that human altruism is different, we need to recognize that this behavior among animals does share something in common with our own. It involves many of the same organic mechanisms that enable us to put the needs of the group over those of the self. The difference is that with us the biological dimension is not the whole story; a complex of biological and non-biological factors is interwoven in human altruistic behaviors.

Let us start by looking at the part of altruism that we do share with other animals. There is an essential biological dimension to the ability of a creature to put the needs of others first; there are specific biological mechanisms that make that possible.

Our understanding of this biological dimension is very recent and is largely the result of new neuro-imaging technologies and advances in biochemical analysis. These new technologies have begun to paint a fascinating picture of the biological and cognitive mechanisms that are involved in altruistic behavior—illuminating both our similarities as well as our differences from other animals.

Vertebrate Selfishness?

While Wilson may have altered his views on the selective forces shaping human evolution, he has not reconsidered the general proposition that vertebrates, because of their competitive reproductive process, must necessarily be "selfish." There is, however, an obvious problem with this outlook—motherhood. Maternal care is unselfish behavior and vertebrates—particularly the mammals—are the very best at it.

The mammalian mother makes innumerable sacrifices for her young. She gives time, energy, and a portion of her own substance to her offspring. Evolutionary anthropologists dismiss this apparent contradiction in the "selfish vertebrate" thesis on the basis that this form of altruism falls within the scope of "kin selection." In providing for her offspring, the mother is simply furthering the fortunes of her own genes. Her generosity towards her young is an expression of genetic selfishness.

That way of looking at the problem, however, fails to address a critical issue. What is the behavioral mechanism that enables the mother to make the sacrifices she does? What enables her to be selfish towards some animals while generous to her offspring? To answer this problem by referring to genes is simply to place a patch over a hole in one's argument. The real question is this: What kind of biological processes do certain genes produce that motivate a mother to perform care-taking behaviors? Once we answer that, we will be in a position to address the issue of whether, and under what circumstances, it may be possible to extend this behavior to a broader circle of community members.

Maternal Love

New research is beginning to reveal the biology of maternal care. It seems that at every stage of the reproductive process—the production of the egg, the implantation of the egg in the uterus, the development of the placenta, and the delivery of the baby itself—the mother's organism produces chemical agents that alter her behavior and that prepare her for taking care of her offspring. These chemicals—hormones, neuropeptides, and neurotransmitters—create a system of rewards and punishments that induces the mother to care for her young.

Donald Pfaff, one of the groundbreaking researchers in this field, has given us an amazing description of the neurochemistry of affection (2007)—and affection is what this is all about. Parental care is not a set of mechanical, robotic acts. It is emotionally driven behavior: behavior that takes pleasure in giving comfort to the object of one's affection.

Using the latest technologies for monitoring activity in the brain (fMRI, PET scans, etc.), researchers have discovered that maternal behavior—nursing, coddling, comforting, grooming, etc.—releases chemicals that activate pleasure centers in the mother's brain. These are the very same centers that are aroused when one eats a good meal or does anything experienced as enjoyable. At the same time, expressions of distress by her infant—whining, crying, and squirming uncomfortably—register in the parts of the brain that process physical pain. Nurturing an infant has a positive physical effect on the mother. It produces a cascade of neurochemicals that reduce tensions and anxieties and that engender a sense of wellbeing.

The neuropeptide oxytocin is central to the workings of this behavioral system. Of all the chemicals produced, oxytocin is "by far the most effective at stimulating high levels of maternal behavior" (Pfaff 2017, p. 101). While it affects many different parts of the organism, perhaps its most important role is in suppressing fear: in breaking down the defensive barriers that normally act to keep animals apart from one another. What we call maternal "love," according to Pfaff, is really a biochemically-induced merger between a mother and her offspring, enabling them to participate in each other's emotional experience.

The biochemistry of parenthood, however, is not entirely restricted to females. In species where the male plays a role in childrearing, the father possesses a chemical that appears to be a derivative of oxytocin: vasopressin. This brings out an interesting side of nurturing behavior. Part of caretaking is protecting one's helpless infants. In line with this, oxytocin has been shown to stimulate aggressive tendencies in the mother. Think Mamma bear. This is even more exaggerated with vasopressin. The male role in nurturing is largely one of defending the family unit.

The implications of this dual nature of nurturing are very intriguing. It would seem that the very biological processes that foster affection and bonding between some, have a tendency to engender hostility and aggression towards others. One cannot help but see the relevance of this to territorial and tribal-type behaviors; the emotional bonds that link an in-group together are easily mobilized to foster aggression against an out-group.

Chemicals Can Only Take Us So Far

This new research gives us an important perspective for thinking about the issues of altruism and cooperation. Pfaff believes the basic chemical-neurological complex that evolved for motherhood and parental behavior provides an essential building block for the construction of higher, more complicated emotional relationships: it "set the stage for the underlying mechanisms of all other friendly and social behaviors" (ibid., p. 80).

This is probably right. Affection—friendship and love—clearly has a biochemical dimension. Breaking emotional connections, as experience teaches us, has painfully-felt, physical consequences. New research has also confirmed that motherhood and parental love in humans make use of the same basic suite of chemical elements (oxytocin, etc.) that are used to foster care-taking behavior in other animals.

However, there are limits to what biochemistry can accomplish. Human relationships are too complex, as Pfaff recognizes, to "be directed simply by hormones" (ibid., p. 109). Human altruism raises problems that cannot be explained in terms of kin selection. Humans do not exhibit altruistic behavior simply towards closely related individuals, but towards groups to which no blood relatives at all might belong. While I believe the biochemical dimension that Pfaff describes is woven into our social behavior, human social dynamics operate primarily on a different plane. Cognitive processes play the principle role in shaping our social connections.

The primacy of intelligence reflects the fact that most human relations are instrumental: they are constructed to achieve certain goals. Whether it is organizing a buffalo hunt, operating a giant industrial conglomerate, or hiring a plumber, humans depend on interactions that are structured by what it takes to achieve certain ends. They are not driven by internal biochemical responses, as are, for instance, the dominance hierarchies of other animals. Human social organization is primarily a cultural rather than a biological fact.

The Cognitive Dimension

The role of intelligence in altruism has been almost entirely focused on the concept of reciprocal altruism. Very intelligent animals, the argument goes, can keep track of their altruistic balance sheets and thus, are able to operate in an arena where give and take can be mutually beneficial.

Wilson also endorses the idea of reciprocal altruism, but he argues that natural selection has pushed beyond this framework. Mutual selfishness cannot produce the cooperative spirit necessary for truly efficient teamwork. Altruistic behavior requires an "irrational" dimension, some kind of emotional response mechanism that can override the self-protective instincts of the individual. This could be something along the lines of the biochemical reward system that operates with motherhood.

I believe Wilson is grappling with an important problem, but he is trying to solve it with too limited a conceptual tool kit (biology). In the next section, I'm going to challenge his model of battling instincts. He, along with almost every other evolutionary anthropologist, simply do not get intelligence; they don't see it as an important factor in generating behavior. Indeed, the word intelligence does not appear in the "Table of Contents" of Wilson's magnus opus—*Sociobiology*—or in its massive, 32-page index.

Wilson sees intelligence as simply a means to achieve instinctive ends. It is, in reality, much more than that; it shapes the ends as well. Intelligence provides a being with the means to make sense of the way events and objects are connected to one another and to it. This determines not only how it understands an experience, but also how it feels about it.

Human intelligence it is not simply animal intelligence scaled up; it possesses a distinctive dimension—a self-reflective capacity. This enables humans to make connections other creatures can't—to experience the world differently. It gives rise to a different understanding of ourselves and our relationship to others. Self-reflection provides us with enhanced empathy and a capacity for greater intimacy—for an emotional depth that is an important part of human altruism.

This conclusion is no longer simply a matter of opinion; the new imaging technologies are beginning to give us concrete evidence of how this works.

Two Different Kinds of Intelligence

In our earlier discussion on reasoning and self-awareness (chapters 4, 5 and 6), I argued that the evidence suggests that the prefrontal cortex plays a special role in higher cognitive functions. All the latest data reinforces that conclusion. At the same time, however, current research is revealing that the pre-frontal cortex is not a single, undifferentiated body, but rather, that it is divided into two major sections that process information differently.

Very early in the game, those performing real-time mental processing experiments (fMRI, etc.) were drawn to a specific region of the prefrontal cortex: the lateral prefrontal cortex (LPFC). It was clear that this area played a central role in problem solving. According to Matthew Lieberman, Director of the Social Cognitive Neuroscience Lab at UCLA, it is "the brain's all-purpose abstract reasoning device (Lieberman 2013)." While all the latest experimental research appears to confirm that the LPFC plays a critical role in our reasoning capacity, at the same time, Lieberman's work has pointed to a fascinating complication.

The LPFC is not the whole story. It comes into play when we try to solve problems that involve a particular kind of "logical" thinking. If we try to work out mathematical equations or think about what would happen if this object were to collide with that one, the LPFC lights up. This, it would seem, is where we solve problems that have to do with physical cause and effect: problems involving the interactions of different objects with one another, or object-object relations.

However, if we were asked to interpret or predict the actions of another person, a different part of the prefrontal cortex lights up. Here the mind is not dealing with "objects," but rather, trying to make sense of "subjects." Subjects don't work by the same cause and effect principles as objects. There is a "logic" to their behavior, but it is not based on physical principles. Subjects are intentional agents and to understand their behavior one must determine their purposes. This, Lieberman tells us, is the job of a different part of the prefrontal cortex: the dorsomedial prefrontal cortex (DMPFC).

The DMPFC lies at the center of a neural network that is specifically focused on making sense of other people's behavior. Lieberman calls this network the "mentalizing system" because the key to understanding the behavior of others lies in figuring out what's on their minds: their intentions. This mentalizing system is almost certainly built upon the incipient "theory of mind" capabilities demonstrated by the great apes. With humans, however, this capacity has grown and taken on new dimensions.

Lieberman presents evidence that the mentalizing system has come to occupy a special place in our mental lives: it has become what he calls the "default setting" of the mind. The mind is always working, even when we are asleep. Sometimes we are problem solving, and our mental energies are focused on one puzzle or another. But there are also times when we like to take a break from serious concentration and just muse or "zone out." At those moments, an area of the brain lights up that occupies more or less the same space as the mentalizing system. This is where our minds like to be.

The human mind, Lieberman concludes, has evolved to make social interaction its top priority. Indeed, we take pleasure in thinking about other people. The evolutionary psychologist Robin Dunbar reports that his research indicates that humans spend almost two thirds of their conversation time talking about other people (1996). Gossip, he concludes, is a fundamental part of the human experience. It serves to sort out our feelings about others, while at the same time creating bonds with those with whom we share our social attitudes and observations.

This makes perfect sense. The most important needs we have, from the time we are born to the moment we die, are all linked to our relations with other people. That's the human bottom line: it's what keeps us alive. Our minds, Lieberman contends, are designed to constantly review and attempt to make sense of our relationships with others. Understanding what makes other people tick is essential for operating effectively within the complex socio-cultural fabric that supports human life.

Social Reasoning

There is one more piece of the cognitive puzzle we need to put into place. That humans should spend a lot of time trying to make sense of what others are up to seems entirely reasonable. It also makes sense that subjects don't act like objects and that we would need a different way of thinking to understand their behavior. But what would that be? What enables us to figure out the goals and intentions of others? What kind of a mental process would make that possible?

One idea that has received a lot of ink recently has to do with "mirror neurons." These neurons, which were first discovered in monkeys, have amazing properties. They fire both when an animal acts and when it observes the same actions being performed by another . When you see someone reach for a glass of water, for instance, the very neurons that would enable you to perform that behavior are triggered in your brain. You can understand what the other person is doing because your brain is actually tracing the movements of their actions. Mirror neurons are clearly part of the picture. They help a creature to understand the movements of another. They undoubtedly play a part in empathetic experience and assist an animal in copying another's behavior. However, Lieberman's experiments indicate their impact is not as far reaching as once believed.

Mirror neurons can help you understand what another being is doing, i.e., reaching for a glass of water. However, they can't tell you why it is performing that behavior: what it intends to do with the water. For that you need to understand a being's motives on a different level.

To get an answer to this problem, Lieberman turns to the philosopher Robert Gordon. Gordon proposes that human beings understand others by projecting themselves into the other's situation. In other words, we attempt to understand the actions of others in terms of what we imagine we would do under similar circumstances. This hypothesis receives experimental support. The DMPFC, the core of the "mentalizing system," overlaps with a section of the brain that lights up when we see ourselves in a mirror, a part that is involved in self-recognition and self-reflection. When we think about others, we simultaneously engage a part of the brain we use to think about ourselves.

Self-Reflection

All of this fits neatly with the analysis of self-reflection dealt with in the earlier sections (chapters 5 and 6). With the emergence of another center of consciousness, a creature develops an awareness of its own mental processes. This gives it a sense of self: it becomes an "object of its own attention." It can reflect on itself: imagine how it appears to others. At the same time, it becomes aware that others have minds and can see things the way it does (theory of mind).

We should take note of the fact that the basis for a "theory of mind" among humans has a dimension that is missing in the great apes. Chimps, because they can experience their own thought processes, can deduce that other chimps must have minds like theirs. Humans, because of cultural tools—signs, and later, symbols—are able to take this to another level. Our knowledge of the minds of others is not just a theory—a deduction. With language, we actually experience the mental operations of others. We know that others think like us because our communication system enables us to exchange our thoughts with them.

Humans, then, experience themselves not just from the "inside," but also from the "outside"—as they appear in the eyes of others. Our sense of self is a construct that meshes both of these perspectives. Psychologists and psychiatrists have always understood this, but it is now possible to place this phenomenon within a broader, bio-evolutionary framework.

The implications of this are profound. A being that includes the perspective of others in the way it experiences itself, while at the same time, looking into itself in order to make sense of the behavior of others, is a being whose experience is, to an important degree, merged with that of other members of its community. This merger creates a framework of shared awareness and a potential for emotional connection that lies beyond anything found elsewhere in the animal world.

A being that is able to see itself like others and others like itself has a unique potential for empathy. This merged experience provides a foundation for an impulse to treat others as you would want them to treat you. As a result of this new experimental evidence, a growing number of researchers are concluding that humans have a built-in predisposition for "fairness." Both Lieberman and Pfaff, for instance, believe that the principle that underlies every universal ethical system—the Golden Rule of "do unto others as you would have them do unto you"—is rooted in empathetic responses that are built into our self-reflective, human experience.

There is, then, an important part of human altruism that should not be looked at simply as one individual sacrificing for another. A being that can feel another's pain, when acting to help another, is on some level, acting to help itself.

Morality

There is one more dimension of altruism we need to look at—morality. We don't normally think of morality as a form of altruism; indeed, morality rarely enters into discussions of altruism. However, as Joshua Greene, the director of the Moral Cognition Lab in Harvard University's Department of Psychology tells us, "the essence of morality is altruism" (Greene 2013, p. 23).

Most works on altruism focus on determining what could motivate individuals to sacrifice for one another. Because individuals seen in isolation are simply organisms, looking at the issue from this individual perspective leads naturally to a search for some biological mechanism that could produce such behavior. And, indeed, new research tools reveal that humans are biologically constructed to perform certain kinds of altruistic behaviors. Besides a built-in capacity for empathy, Greene cites experimental evidence that humans receive a biochemical reward for acts of generosity; we get a shot of oxytocin when we tell a stranger where the subway station can be found.

It makes sense that a species that depends on cooperation would have evolved a biology, as E.O. Wilson argued, that would incentivize altruistic behavior. Morality, however, cannot be explained in biological terms. It is not just about empathy or generosity; it's about playing by the rules. Generosity is a sacrifice that one person makes to benefit another. Morality, on the other hand, is a not about giving to another; it's about doing what is right. That's different. Morality is not experienced as a matter of generosity but rather as a matter of duty. It is not really about empathy or generosity; it is about right and wrong.

Morality takes the discussion of altruism from an individual or biological level to a cultural level.

Morality is Cultural

Morality is a form of cultural behavior. That means it involves a cultural device—the use of something that has an existence that is independent of our biology. This dimension is generally overlooked in discussion about morality.

While the evidence suggests that humans possess certain biological tendencies toward the performance of altruistic behaviors, looking at morality as a self-sacrificing instinct contains an obvious flaw. The reason human communities develop moral rules is precisely to restrain individuals from acting on their spontaneous self-directed impulses.

Yes, we have empathy and generosity, but we also have emnity and selfishness. Human social life is rife with conflict; the ability to see a rival as the same as oneself is not as easy as the saintlier among us might suggest. Fear and anger distort our perceptions, and humans are prone to demonize those they perceive as some kind of threat. Ignorance and prejudice can override a sense of fairness—and the possibility of violence is present in every clash of wills. And then there is just good old-fashioned lust—a craving for objects, attention, power and sex. Humans may have a tendency towards empathy and generosity, but we also have a tendency towards violence and taking more than we give

Morality is a set of rules that place the needs of the society above the emotionally driven and self-directed impulses of individuals. It is an essential requirement of a human way of life.

Freud?

The idea that morality emerged as a way of curbing biological instincts was the signature idea of Sigmund Freud—and it dominated thinking during the greater part of the twentieth century. In Freud's model, culture was depicted as a thin veneer of restraints resting upon a violent and selfish biological core. That's not an accurate view of the issue.

Culture and human biology are not, as Freud believed, two different things that have become superficially pasted together. They evolved in association with one another, as two interconnected dimensions of a single life process. Human biology is not at war with culture. At the same time, however, the two are not identical. Culture is not a direct expression of human biology; it consists of things that are independent of the human organism and that endow it with capacities that are not built into it.

Human intelligence is perhaps the clearest example of how this works. Our cognitive capacities have an essential biological dimension—our huge, over-sized brain with its built-in self-reflective capacity. However, our incomparable mental achievements are not traceable to that biological organ alone. A human level of intelligence is the product of a merger of the brain with a cultural tool: language. This merger carries human intelligence beyond the framework of the biological brain.

Human social capacities are the product of a similar kind of merger. The ability of humans to live in complex, ever changing, cooperative communities is largely due to their ability to formulate and assimilate moral codes. A moral code is a cultural device—a set of rules that enables individuals to elevate their social behavior from a level that is grounded in their own individual needs and desires to one that is based on the requirements of collective life.

Right and Wrong: A Higher Principle

From an anthropological perspective, I have always been intrigued by the story of Genesis—the origin myth that appears in the ancient Hebrew Bible. The narrative begins with Adam and Eve eating the fruit of the Tree of Knowledge. The knowledge they acquire is not about the workings of the world; it is, rather, an ability to distinguish between right and wrong. It is this awakening to morality that the early Israelites saw as the event that placed humanity outside the framework of the natural world.

What exactly is right and wrong? What is so special about these moral categories that they are held to represent the divide between humanity and the animal world?

Animals know only the needs and desires of the individual: even the maternal love of the mammalian mother is an expression of organic impulse. Right and wrong, however, exist on different planes; they do not pertain directly to the needs of individuals but to the needs of a community. They are the rules that individuals must follow if the community is to function as a collective enterprise.

Right and wrong, then, represent something "higher"—something that stands "above" individual impulse and preference. Without adherence to these higher rules, that is, without some check on individual impulse, society could not function as an organized body—and human cultural life could not exist.

The Power of Morality

While morality represents the requirements of the larger community, its power to affect the behavior of the individual derives not only from external pressures, that is, not only from pressures exerted by other members of the community. Individuals are taught moral principles from an early age, and they experience them not only as right and wrong, but also as good and bad—as the basis of a community's approval or disapproval. These rules become incorporated by the individual into their own self-evaluation—their own sense of self. Our sense of self, then, is inseparably connected to the moral standards of our community. Moral principles become a sort of social instinct—a moral compass that guides us from the inside. When we violate those principles, we feel—or should feel—shame and guilt.

Morality, then, like human intelligence, is the product of a merger of biology and culture. It enables people to operate by rules that are not built into their biology, but that become incorporated into their cognitive and emotional systems, and which are essential for functioning in a complex cultural community.

Chapter 16

Conclusion

The Emergence of a Cultural Being

In the beginning of this book, I stated that we haven't really known what kind of being we are—how we fit into the natural world. Are we an animal or are we something else? And if we are something else, what could that something else possibly be?

At that stage of the discussion, we lacked the cognitive tools to provide a clear answer to that question. My goal in this book has been to establish that we are not simply another kind of animal but that we are, indeed, something else. However, when I began this effort, there was no model I could use to describe what that something else might be. That's what I have sought to do here: to construct a concept—to build a model—of the unique kind of being that we are.

Looked at as individuals, we humans are clearly a kind of biological organism. Our biology could best be described as that of an evolved, terrestrial great ape. This would seem to put us in the animal camp. At the same time, however, we are not like any other animal. When Darwin developed his theory of evolution, he thought of animals as beings whose behavior was a function of their biology. Their different capacities were seen as expressions of their different organic faculties. Birds are constructed for flight, fish for swimming, and great apes for life in the forest canopy. As Darwin saw it, every life form was a product of natural selection—of an evolutionary process that shaped its organism to cope with the particular challenges posed by its way of life. On a general level, he was right.

However, humans are different. Our evolution shaped us to relate to nature differently from any other creature. Indeed, the principal difficulty in answering the question "what are we?" was not in assembling all the latest evidence about the evolutionary path our ancestors followed. It was, rather, in finding a way out of a conceptual trap that left no room for something truly different to emerge from an evolutionary process.

The field of human evolution has operated under the assumption that humans, as products of the same kind of evolutionary process as every other animal, must work by the same rules they do. That assumption has been a major stumbling block in efforts to develop a credible model of human evolution. The logic appears to be unimpeachable. If it were true, that we are the products of the same kind of evolutionary process as every other creature, then we would, on a fundamental level, be the same as they are. But the assumption is flawed. It ignores the unique features that gave our species' evolutionary process its distinctive trajectory. We could not have become the exceptional beings that we are unless there was something importantly different about our evolutionary process.

However, the proposition that humans are the product of a different kind of evolutionary process is, to say the least, very controversial and would be rejected by the overwhelming majority of evolutionary scientists. On one level they would be right. There is only one kind of evolutionary process: the one discovered by Darwin, adaptation by means of natural selection. Didn't we also adapt by means of a natural selection? What has made human evolution so difficult to understand is that the answer to that question is both yes and no.

A Unique Evolutionary Process

Human evolution was not just another example of some creature adapting to its environment. To begin with, our great ape ancestors were not just another kind of animal. They represented something new—the emergence of a new intellectual plateau within the animal world. They, along with a small handful of other higher vertebrates, had developed the germs of a self-reflective intelligence: an ability to "think about" their thoughts, that is, to re-process their perceptual models. This reflective capacity provided them with the germs of self-awareness and enabled them to construct mental models that allowed them to perceive how different objects are related to one another—cause and effect.

This more highly evolved intelligence opened the door to new kinds of behavior. We see an expression of this in the capacity of contemporary great apes to use tools. When an animal uses a tool, it is using something other than itself—other than its own organism—to affect objects in its environment. It is acquiring a capacity that is not built into its biology.

The ability to perform this kind of behavior was critical for our australopith ancestors. They occupied an environment that was changing faster than their slow-moving evolutionary processes could keep pace with. Their biology had evolved in a set of conditions that were increasingly at odds with their new circumstances. Most animals when faced with this kind of situation would simply have gone extinct. But the intelligence of these great apes, along with their facile hands and upright posture, gave them a different option; they could acquire new capacities by making use of external objects.

This new behavior almost certainly began with very simple hand tools. Digging sticks and stone flakes made up for the absence of claws and canine teeth. As their environmental conditions continued to deteriorate, they needed to extend this principle into other realms; they began to create other kinds of "tools." They discovered how to make communicative signals—signs (gestures)—that could represent objects and events in their environment and which enabled them to share information at a level no other creature could. This made it possible to collectively determine goals and to coordinate their actions, combining their energies in unprecedented ways. All of these behaviors—tools, signs, organization—are ways of creating capacities that do not depend on biological changes, but rather, on making use of the properties of things that lie outside the organism

Darwin defined evolution as "adaptation by means of natural selection." He saw evolution as the inevitable product of a feedback loop between an animal and its environment. The animal's reproductive process would inevitably produce novelties, and the environment would determine which of those novelties would be rewarded through natural selection. That's not all that was going on with our early ancestors; they were not simply adapting to their environment with their biology. Our great ape ancestors were discovering ways to acquire capacities that did not depend on changes in their biology. This was a new behavioral strategy and it represented a new stage in the development of life on our planet. It marked the appearance of culture. A tool, by giving a creature a capacity that is not part of its organism, enables that creature to adapt without having to undergo evolutionary changes. Culture is an alternative to natural selection.

What makes all of this so complicated is that while culture may be an alternative to natural selection, humans could only have evolved a capacity for culture by means of natural selection. Cultural behavior has its own biological requirements. The biology of our great ape ancestors had not evolved to make use of cultural devices, but rather, to solve an entirely different set of problems. Natural selection had to reshape the great ape organism to enable it to perform effective cultural behaviors.

This new cultural survival strategy altered the way natural selection operated. Instead of working to provide our ancestors with the biological means to cope with the challenges they faced, natural selection rewarded modifications that enhanced their cultural capacities—that provided them with alternatives to biological-based solutions. Our species, then, did not adapt directly to its environment; it adapted to its unique way of adapting. It adapted to culture.

How Different Are We?

The fact that we are the products of a unique evolutionary process does not, in and of itself, signify that humans are something other than a different kind of animal. Couldn't we be culture-bearing animals? I believe that would have been a good description of our hominin ancestors throughout much of our species' evolutionary history. But it's not where we ended up.

The evidence suggests there were two major phases of human evolution. The first consisted of the transformation of a great ape—an australopithecine—into a being whose way of life depended on cultural skills—*Homo erectus*. *Homo erectus* had adapted to culture in three critical ways. The most important was that its brain had doubled in size from that of its australopith ancestors. Intelligence is the principal prerequisite for cultural behavior. The ability to use things outside one's own organism to acquire new capacities involves a higher order of behavioral complexity.

Cultural behavior is mediated, that is, a whole complex of intermediate steps lies between the actor and his or her ability to accomplish some goal. To be able to cooperate, for instance, our early ancestors would have needed to use a system of signs to communicate a plan—an action sequence based on coordinating the roles of a number of different actors. That lies way beyond the mental capacity of any other animal. The complexity of this kind of behavior is the reason that humans ended up with brains that are seven times larger than those of the average mammal, e.g., dogs, horses, or lions.

The second important way that *Homo erectus* had adapted to culture was with its anatomy. A great ape's body is not a good vehicle for cultural skills. The kinds of hands and arms required for the effective manipulation of objects demanded changes that were incompatible with our lineage's ancient arboreal way of life. In order to make those changes, the early hominins had to abandon the trees and be able to hold their own on the floor of the savanna.

Homo erectus' third adaptation was what we might term social-reproductive. The larger brains selected for by culture brought with them a host of difficult problems, among them longer pregnancies, increased infant dependency, and extended childhoods. It was absolutely essential for a species that was becoming increasingly reliant on meat eating and hunting to find a way to support childbearing females. This would have involved food sharing and, almost certainly, some kind of sex-based division of labor. *Homo erectus*, the evidence suggests, had met this requirement.

Homo erectus had evolved the mental, anatomical, and social-reproductive essentials of a new, culture-based way of life. It survived by participating in a community that used tools, could communicate with signs and mime, and that could cooperate at a level no other animal could. Was it an animal—or had it become something else? It was, I believe, a halfway creature: different from any animal we are familiar with, and most likely recognizably human in many ways. Yet, on an important level, it still remained within the behavioral sphere of the animal world.

Its culture was very primitive—no language, no fire, no art or ritual, and it could not even haft a stone point to a tool or weapon. However, the most telling fact was that for about 700,000 years, that is, throughout the period when it was the dominant hominin, there is almost no evidence of novelty or cultural change. That would await the appearance of a new, larger-brained hominin. At this stage of human evolution, cultural development was still inseparably tied to changes in biology, to evolutionary processes. *Homo erectus*' capacity for cultural innovation, then, had an upper limit that was determined by its biology. On this critical level, it still operated within the framework of the animal world.

Homo sapiens

The real break with the animal world came during the last 120,000 years of our species' three-million-year evolutionary process. This, I believe, is when a fully developed language capacity came into being.

Language, like every other aspect of culture, is based upon the use of external tools: symbols—a symbolic code. This development most likely grew out of a switchover from visually based signs (gestures) to auditory signals (vocalizations). The increased use of sounds, which have certain communicative advantages over signs, would eventually have had profound consequences. Most sounds, as opposed to signs, have no perceptual relationship to the things they stand for, and thus have an inherently symbolic character. It was the evolution of an ability to encode experience into symbols that was the game changer.

Symbols are importantly different from other cultural tools, i.e., technology and social organization. They are not a force in their own right, that is, they do not directly add to an individual's physical powers. Nevertheless, they are the most powerful instruments we possess. They provide the mind with tools that allow it to construct abstract mental models. They enable humans to transform perceptions into concepts—into mental models that operate on a different level from imagery. When that little creature over there is labeled mammal, it is placed in an abstract category and elevated to a level that brings to mind not only its specific qualities, but what it shares in common with a multitude of different animals—bats, whales, elephants, and mice.

Concepts, as opposed to perceptual models, are not automatically formed by the biology of the brain. As we saw from Vygotsky's work, they must be constructed linguistically through conscious effort.

Language allows humans to build mental models that do not correspond to anything we directly experience with our senses. This is what the philosophers called knowledge. It enables humans to represent experience at a higher level of generality, to package information in new ways. The term animal, for instance, provides humans with an abstract category—a "folder"—that enables us to organize data about the animal world at different levels of generality, such as species, genus, family, class, order, phylum, etc.

The ability to construct abstract models enables humans to discover patterns and relationships that are not revealed at a sensory level. The distillation of general principles—logical, mathematical, or physical—enables the exploration of realms that lie far beyond the reach of our perceptual powers. They allow us to enter into the world of sub-atomic particles on one hand, and the dynamics of galaxies on the other.

Most of the breakthroughs in humanity's understanding of the world have been based on the discovery of some new general principle, some conceptual innovation that has enabled things to be seen in a different light. This is what an "open-ended intelligence" is really all about. Symbols (language), by permitting the construction of more abstract and inclusive categories, gave birth to an intelligence that could build on itself and develop means for incorporating an ever-wider range of experience into ever-more adequate models of the world.

A Collective Consciousness

Language affected human intelligence in two important ways. On the one hand, as I argued above, it provided tools (symbols) that the mind can use to construct and represent concepts. But human mental powers cannot be understood simply as the products of individual mental capacities. Of equal importance is the fact that language gives rise to a collectively organized intelligence.

The key to the power of the symbol is that it has a real, material dimension. It is not just a synapsis in an internal neurological process. It has an objective form; it exists as a sound or a sign, something that is independent of the human organism. This is what enables it to be shared, to serve as an instrument of communication.

Words don't belong to individuals. A language is the property of a community. The words we think with, the concepts that elevate our understanding, are not our own individual inventions, but are the distillates of a community's collective experiences. Everything we know other than what is given to us directly by our sense, is built upon an information base that has been encoded into our language over the course of centuries.

Language produces a being that is simultaneously the most autonomous and the most collective of all life forms. No other creature has the capacity to represent, reflect on, and modify its inner mental processes the way that humans can. At the same time, no other creature constructs its awareness out of shared mental models. When we think, we think with a linguistically constructed consciousness. Language transformed human intelligence into an open-ended, social/historical phenomenon.

The Culturally Constructed Community

While every culture rests on a linguistically coded, conceptually organized system of knowledge, a human community is much more than a set of ideas. It is an organized life-support system. Every human community must address the biological needs of its members. It must provide food, shelter, security, and the opportunity to biologically reproduce. That's the bottom line. How it does this is what distinguishes us from all other living creatures.

The main thesis of this book is that humans survive by making use of objects and forces that are not part of our organisms. These objects, it is important to understand, have properties of their own. They can only serve our purposes if we accommodate their independent requirements.

If a community is to make use of a certain technology, for instance, it must organize itself to make that possible. When the early Sumerian farmers began to move into the fertile Mesopotamian flood plain, they had to learn how to control water. They needed levees to protect their crops during the seasonal floods and irrigation canals to transport water during the dry periods. Moving immense amounts of earth required large numbers of people working together—organization. Organization, in its turn, put a premium on planning, and appears to have gone hand in hand with the emergence of organizational specialists—an authority structure.

Life on the flood plain posed a number of difficult problems. While there was plenty of soil and water, there was little else. Lacking wood and stone, these communities had to organize themselves for large-scale, long-distance travel and trade. Trade and complex organization require records, and not surprisingly, the Sumerians were the first people to devise a writing system (3,300 BCE). This was a cumbersome affair that involved learning hundreds of different pictograms and characters, and it demanded specialists—scribes, and schools for training them. And so it went, each development building on and demanding another, eventually resulting in a level of cultural complexity we call civilization.

A human community is a construct, a synthesis of biology and culture. Every cultural innovation—technological, organizational, or symbolic-conceptual—requires an alteration in human behavior and demands changes in the form and functioning of the community. There is nothing in human behavior that can be traced entirely to biology; nor, for that matter, that can be entirely distinguished from it. Even nutrition—what we eat, how we prepare it, and the way we consume it—is embedded in cultural norms. This is true for sex and reproduction as well. With humans, reproduction is bound up with a complex set of rights and obligations. Every human community establishes rules that dictate who is sexually available, who has obligations to the child and the mother, and to whom (kin, tribe, nation, etc.) the child belongs.

Anthropology has performed a valuable service in documenting the incredible variety of different ways that people have merged their biological and cultural sides into a coherent way of life.

From Evolution to History

The emergence of language gave birth to an organized, culturally constructed community and to an open-ended, collective intelligence—to a new stage of human development. Humans were now able to create new cultural capacities without having to alter their biology. At that point, development was no longer bound up with evolutionary processes, but rather, depended on cultural innovation. Biological evolution passed over into historical development.

This transition from biological to historical development appears to have reached an important threshold around 70,000 years ago. While the biology of humans continues to evolve in certain ways—where survival is affected by processes that lie beyond culture's protective umbrella (pigmentation, immune responses, size, etc.)—it appears that the basic linguistic and intellectual foundations of a cultural way of life crystallized in a lineage of *Homo sapiens* that emerged in southeast Africa around that time.

For the first 50 or 60,000 years of historical development, cultural innovation took place within Stone Age parameters. While there was technological progress during this period, we don't see any dramatic changes until around 12,000 years ago. That was when the last Ice Age came to a close, and far-reaching ecological changes demanded that humans re-tool their cultures—in particular, that they abandon big-game hunting and make use of new sources of nutrition. This is when people discovered how to domesticate plants and animals—when they gained control over their food supply.

Agriculture was transformative. Population multiplied at an unprecedented rate and people could now settle in one place. They built houses and villages, and began to construct ever larger and more complex centers of collective life. These changes were made possible by a wide range of cultural innovations. Improvements in food production were critical—new crops, irrigation systems, ploughs, draft animals, crop rotation, etc. These advances went hand in hand with a wide variety of technological advancements—ceramics, kilns, metallurgy (metal weapons and tools), wheeled vehicles, etc.

What is not generally understood is that after the discovery of agriculture, the most significant inventions of the ancient world took place in the social realm. As population outgrew the old tribal organizational forms, new social hierarchies emerged, and along with them, new political, religious, and military institutions. This is when rules and laws came to replace norms and traditions.

Organization, the ability to employ large numbers of people effectively, was the principal cultural advance during this early period. It was the emergence of centralized political power—states—that enabled the monumental achievements we associate with the emergence of civilization (pyramids, ziggurats, massive canals and irrigation works—and organized military forces).

There is a tendency to think of history as a narrative created by extraordinary individuals and events. Those unique factors do play a role in shaping the world we live in, but what we think of as history is better understood as the growth process of cultural communities.

The history of Rome cannot be told through the lives of Caesar, Augustus, or Constantine. Those men had their impact, but Rome was there before and after them. Rome was a cultural community—an organized construct that evolved through a series of stages as it sought to consolidate its ever-growing demographic and organizational supremacy over its neighbors. While Caesar was clearly a military genius, it was Roman culture that provided him with the power he was able to use with such effect.

So—What Are We?

How are we to make sense of all of this? What kind of being are we? How do we fit into the natural world?

Before attempting to say what we are, we should begin by clarifying what we are not: we are not just another kind of animal. This is a difficult idea to accept because we share something fundamental with every animal; like them, every human individual is a biological organism. Nevertheless, it is critical to find a term that can distinguish us from the animal world. This is not because there is some shame in being a part of that world, but because the word animal denotes a life form that represents a particular stage in the evolution of the organic world. To use the term animal to describe humanity is to fail to recognize our defining characteristic: that we are a life form that has moved beyond the framework of biology.

This is not the first time that life has gone through a dramatic transformation. Indeed, the whole history of life on this planet has been one of continuous transformation. Beginning with the consolidation of certain inanimate chemical elements into sentient, self-reproducing life forms, we see a continuous sequence of new kinds of biological organisms: archaebacteria, protozoans, multicellular forms, fungi, vertebrates (i.e., fish, amphibians, reptiles, birds, and mammals). Those that came later possessed features that enabled them to go beyond the limits of their predecessors.

Humanity represents one of these transformational steps. We are a new kind of life form. In this book, I traced the process by which an animal, a kind of great ape, evolved into a being whose life process extends beyond the framework of biological organization. Beginning with simple tools, our ancestors' survival came to depend on the ability to assimilate the properties of external objects and forces—a process that concluded with the incorporation of our cultural inventions into our innermost core. With language, we do not simply use a tool or external object to gain a capacity; we integrate symbols into the workings of the brain, and by so doing, are able to transcend the limitations of an organic-based intelligence. The emergence of a language-based, collective intelligence gave birth to a new kind of collective life-force—the cultural community. We are, then, a being that cannot be defined on an organic level. We are a cultural life-form.

Culture represents a new stage in the development of life; it came into being when the life-impulse moved beyond the limitations of biological form. Biological organisms, it would seem, can only develop to a certain point: they can only encompass a certain level of complexity. On the other hand, the development of culture-based capacities, both physical and mental, would seem to be limitless.

However, before we revel in the possibilities of limitless development, we need to get past the daunting challenges that nature has currently placed in our path.

Global Warming and Human Evolution

There is a gap—a disconnect—between humanity and the rest of nature. This gap is not momentary or accidental; it is a fundamental dimension of what we are. Culture is an indirect way of solving problems. We don't dig, butcher, communicate or even think with our organic inheritance alone. Instead, we rely on our cultural constructs, the powers and qualities we acquire through assimilating the properties of things outside ourselves. Our relationship with nature is mediated; culture occupies the space between our organisms and the surrounding world.

Recognizing this indirect relationship provided the key for understanding the dynamics of our unique evolutionary process. Our biology adapted to culture rather than directly to the external world. But as is now painfully clear, the gap that culture creates between our organisms and the natural world harbors an existential threat. While culture has placed us beyond the framework of ecological checks and balances that operate for other species, it cannot change the fact that, like every other life form, we ultimately depend upon the resources of our planet. The problem we face today is that while our biology has adapted to culture, our culture has not yet adapted to nature. Humanity has not yet completed its developmental process.

Two Different Phases of Development

At first glance, human biological evolution and historical development would seem to be two entirely different processes. And on one level, as anthropology correctly argued, they are. However, when we step back and look at the big picture, it becomes possible to recognize that they are also connected; that they represent two different phases of a single developmental process.

The first phase was biological. This was set in motion when our australopith ancestors began to rely on simple cultural devices. That altered the forces of natural selection and gave rise to an evolutionary process that produced a biology that was constructed to perform cultural behaviors. This biological phase culminated some 70,000 years ago with the evolution of a symbolic language capacity. At that point, biological evolution no longer played an important role in the human adaptive process. However, this did not represent the end of human development, but rather, the beginning of the second phase.

Why should human historical development be considered a second phase? What does it have to do with our evolutionary process? Didn't the emergence of culture break the link between biological and historical development? Yes and no. Our evolutionary process produced a being that is biologically predisposed to accumulate information—to develop its culture. With language, we evolved the capacity to collectively accumulate and advance knowledge, to construct and share a system of ever-more adequate models of the world.

However, while we emerged from our evolutionary process fully equipped to begin learning about the world, we were born in ignorance. There is an important difference between a biological potential to accumulate information and the actual amount of information a being possesses at any given moment. The former is given at birth, while the latter is the product of a learning process.

Our early *Homo sapiens* ancestors had the same brain, the same biological "hardware" we do. However, human intelligence, seen as the ability to understand and solve the problems we face at any given moment, is not a biological given. Our actual intellectual capabilities are primarily artifacts of culture. They depend upon the state of the cultural "software" that is the collective property of a community.

When our early *Homo sapiens* ancestors evolved a capacity to construct conceptual models of the world and stepped beyond the cognitive limits of the animal world, they began that process equipped only with the information that their senses could provide for them. While they had an extensive practical knowledge of their environment, they knew nothing about its inner-workings—how things became what they are and why they behave as they do. They imagined that the natural world was made up of living, intentional forces, and that it operated by the same principles that humans do. It is only now, after 70,000 years of cultural development, that we possess a body of knowledge that enables us to have a workable understanding of how the world came to be the way it is and how we could have emerged from it.

The crisis we face today is not an expression of an irreconcilable opposition between humanity and the natural world. It represents, rather, a stage in the development of our consciousness—a period when we did not understand our integral connection to the natural world. We saw ourselves as outsiders. Nature was viewed as simply a means for our ends and we blindly pushed our cultural powers ahead without any awareness of their impact.

The Importance of Science

The picture that emerges from this work is of a humanity on the brink of a self awakening. I believe that is fundamentally what is taking place. But besides all the "objective" problems we face, that is, all the hard work it will take to heal a badly scarred nature, there is another problem we have to confront: disinformation.

A cultural being is dependent on the information it possesses. It is our knowledge of how nature works that has enabled us to use its properties for our own ends. Knowledge, as I have argued, is not given to us directly in experience; our concepts, our models of things, must be constructed. This is where science comes in. Science is a cultural tool, a set of rules for processing information that is vital for the construction of knowledge.

The heart of science is the experimental method. To qualify as knowledge, an idea must be confirmed by empirical evidence—by something that is independent of the internal logic of the idea itself. Such evidence is the only real test of its validity. Science cannot guarantee truth, but it is an invaluable tool for distinguishing grounded beliefs from groundless speculation.

One of the problems we face today is that the information we need to heal the breach with nature can only be obtained through scientific work that is far removed from everyday experience. We are not just dependent on science, but also on scientists—on individuals with the special skills and training needed to understand the processes that are threatening our world. This reliance on specialists would not be a major problem if there were a collective consensus about the need to better understand how human behavior is negatively impacting our planet. But currently that is not the case. The gap between the information we are receiving from the sciences and everyday experience has opened the door to sophisticated disinformation campaigns,

Humanity has a spotty record when it comes to dealing with new information. On one hand, it is often enthusiastically welcomed; it adds to the store of human powers. But at the same time, new information can affect the status quo, alter previous relations and reverse the fortunes of different actors. Whether by Roman crucifixion or the fires of the Inquisition, those whose interests have been jeopardized by new ideas have struggled mightily to staunch their spread. While naked force still plays a role in many parts of the world, the battle against new information takes on a more subtle form in the more highly developed centers of culture.

In America, we have witnessed the growth of an entire industry devoted to undermining public confidence in scientific research. David Michaels (2020), drawing on his long tenure in the Occupational Health and Safety Administration, documents the growth of a movement initiated in the 1950s by the tobacco industry, and that spread from there to the producers of pesticides and toxic chemicals, the automobile industry, the oil and petrochemical industries, multinational pharmaceutical companies, and a host of other major players in our economy. Billions of dollars are spent every year to create doubts about the claims of science and the integrity of scientists, particularly in relationship to climate change.

Humans are vulnerable to attacks on their knowledge base. The great bulk of what we know comes from others, from our culture rather than from our own direct experience. While this collective cultural consciousness is one of our great strengths, it has its down side. The beliefs of others, the general acceptance of certain critical assumptions, plays an important role in sustaining our own belief systems. When falsehoods are manufactured and repeated over and over again (The Big Lie), even baseless nonsense can begin to acquire an aura of plausibility. In America, the attacks on science have reached a point where truth is increasingly measured by who says what and which side one is on.

This is a frightening development, and the confusion and uncertainty it engenders feeds an impulse to seek security in the sham embrace of authoritarianism.

The Path Ahead

This book ends where our current problems begin. It would take another book to address those issues—and a different author. But I would like to end this work by calling attention to the amazing nature of the creature whose genesis I have documented in these pages.

We are a unique kind of life-form—a self-creating being. That may sound more like New Age fantasy than hard-nosed scientific analysis. But it's not. It's what this narrative has revealed us to be. There is no doubt that the road ahead will not be easy, and there are many who despair of our ability to overcome our deep divisions and our selfish, violent and irrational impulses. If it were human nature to be the way we are today, then, indeed, our future would look bleak. But what we are today is not all that we can be.

We truly are a self-creating being. That's not hyperbole. We assimilate the properties of external objects and forces, and through our own creative efforts, transform them into new personal capacities. That is how we move beyond what we are born with—beyond our biology.

To go beyond biology, however, does not imply that we have severed all connection to the animal world. As individuals, we are flesh and blood creatures with roots that are deeply embedded in that world. We will always possess a life force that is needy, impulsive and not entirely subject to rational control. That dimension will always be with us, an endless source of drama—of joy and tragedy.

But we possess another dimension. We alone of all the creatures on this planet can think about what we are. Even more importantly we can construct a model of what we would like to be. We are a free being, not free to be whatever we want to be, but free to choose from those options that history makes available. If we are to survive our current existential challenge, it will be because we make a critically important choice—to join forces in common effort. We are not being asked to place the good of the community over that of the individual, but simply to recognize that they are inseparably connected.

The View from the Cosmos

I began this book by asking the question "what are we"? Are we an animal, or are we something else? I ended by concluding that we are not an animal, but something else—a new kind of life-form, a cultural being.

In closing, I would like to pose a different question. What is a cultural life-form? This question shifts the focus from our relationship to the animal world to a much larger issue. What does our genesis tell us about this incredible universe that gave birth to us?

The evolutionist/philosopher Aldous Huxley proposed that humanity, when placed within the context of the development of the Cosmos, should be seen as the vehicle by which nature is becoming conscious of itself. My first reaction to this idea was that it was preposterous—way over the top, entirely too mystical. However, on reflection, it becomes clear that there is something to this. After all, we are a part of nature—we grew out of the animal world. And we are also ceaselessly increasing our knowledge of the origins and workings of the universe. We now know about the Big Bang, the formation of galaxies, and the genesis of our own solar system. We have learned about the consolidation of the earth and when and how the moon was born. We can trace the emergence of primitive life forms back some 3.7 billion years—and we are coming to understand the process by which they eventually grew into us. We are irretrievably on a path towards constructing a unified conceptualization of the universe.

I am fascinated by this big view of humanity, but it's a bit too teleological for my taste. It suggests that we have come into being to fulfill a function: that where we ended up explains how we got here.

That clashes with my science-based, cause and effect world view.

I am more comfortable with a bit more prosaic vision of the big picture. Like Huxley, I don't think our emergence is entirely accidental. I believe there are other intelligent life forms in the universe, and that like us, they grew out of the evolutionary logic of their planet's biosphere. Life naturally expands, pushing the boundaries. As it evolves, it moves into empty spaces; in the words of the classical philosophers, it abhors a vacuum. Developing first in water, early life forms eventually invaded dry land, and from there, moved into the air. Behaviorally, evolution gave rise to increasingly intelligent forms, and eventually to a species that moved into a new void, a way of life made possible by a higher intelligence. That, I believe, is what culture is—what we are.

Whichever of these "big" visions turns out to be true (the philosopher Hegel argued that the great mystery of the universe is that both are true), I take away from this work a sense of awe I had not previously possessed. We are amazing beings, the products of an amazing evolutionary process, generated by an amazing universe. While we need to appreciate ourselves more than is the current fancy, we have to grow up fast and stop thrashing about in the dark.

Intelligence is an awesome force; if it is wielded blindly in service of narrow, immediate aims, it becomes destructive. An intelligent being only has a future if it can come to understand the responsibilities that go with its unique gift. That's what higher purpose is all about; in pursuing our own ends, we must simultaneously nurture and preserve nature's life-giving forces.

APPENDIX 1

EVOLUTIONARY TIMELINE

25,000 YA	Last Neanderthal
60,000 YA	Contemporary *Homo sapiens*
300,000 Y	Archaic *Homo sapiens* Neanderthals—Denisovans
800,000 YA	*Homo heidelbergensis*
1.75 MYA	*Homo erectus*
2.0 MYA	*Homo ergaster*
2.6 MYA	*Homo habilis*
3.4 Million Years Ago	*Australopithecus afarensis*

APPENDIX 2

A word about my use of philosophy as a source of understanding about intelligence.

Far and away the most difficult and time-consuming part of this book was the labor that went into extracting a model of intelligence from the works of Kant and Hegel. I do not have formal training in philosophy, and it took me years to understand the terms and assumptions that are integral to their works. My biggest problem, however, was not just a lack of familiarity with the subject matter, but rather, the fact that Western philosophy's model of the mind possesses a critical flaw. It does not recognize the role of language in the construction of conceptual thought. Because language is unique to humans, if one can't distinguish the effects of language from the operations of the organic brain, it becomes impossible to find a connection between the intelligence of humans and that of other animals.

To be able to use Western philosophy's model of the mind to analyze the cognitive difference between the *Cebus* and the chimp, I had to get "beneath" language; I had to translate the model of logic that the philosophers used to make sense of language-based discourse into a model of logic that could make sense of behavior—of activity.
If one reads Kant, one will look in vain to find the model that I ascribe to him in chapter 4. But while that model might not be there in a readily recognizable form, the basic elements of it are, in fact, present in the works of both Kant's and Hegel. Logic in Kant produces a system, and in Hegel, a process.

To solve a complex tool-using problem, the chimp had to be able to represent its environment as a system and its own behavior as a process. This insight opened the door to pursuing the connection between Western philosophy's model of the mind and the intelligence of other animals. It led, eventually, to the recognition of the role of language (symbols) in elevating human intelligence to another level.

APPENDIX 3

For our purposes in this book, the differences between Hegel and Kant are not significant. But in the bigger landscape of Western philosophy, the differences are very important.

Kant saw the mind as built to organize information into systems—to give it a logically coherent form. This was the work of what Kant called the Understanding.

Hegel agreed with this, but he argued that the construction of a systematic model was not the finished product, but rather, a stage in the process of coming to know something. The mind advances its understanding by creating systematic models—bringing a body of information together in a unified way, by means of a single principle.

He saw that as an essential step in our learning process. But such a step necessarily fails to include information that is at variance with the principle that is used to organize these models. Having taken this step and having advanced our understanding, we then begin to see the shortcomings in the model—contradictions that were brushed aside in the earlier synthesis.

Eventually, the contradictions begin to surface, and the model starts to break down. The old synthesis no longer makes sense— it can no longer adequately account for the reality it previously appeared to explain. Then a new process begins. We seek to find a new synthesis—a way to incorporate the contradictions in the old model into a new system. The new system, however, will have the same fate as the old one. It will eventually be shown to fall short of reality, and need to evolve in order to incorporate the contradictions that will become apparent in it.

Kant drew an indelible line between the models the mind constructs and the true nature of the external world. He concluded that we can only know what our models tell us about the world, we can never know what the world is in and of itself. Hegel accepted this on one level; but he saw our models of the world as forever evolving—incorporating ever-more information about the external world.

Hegel's philosophy is opposed to all forms or dogmatism. For him, knowledge is an eternally developing process. Kant performed a service by throwing light on the process, but he mistook a step in the process for the process itself.

APPENDIX 4

Mythology and Philosophy

During the course of my research on early modern philosophy, I became interested in the origins of philosophy as an intellectual tradition. When did people begin to ask the questions we associate with philosophy? Is there anything different about it, or is it simply an expression of humanity's general need to speculate about the nature of things?

The Ancient Greeks themselves saw philosophy as something radically new and different. Indeed, they traced its beginnings to Thales of Miletus who, in the early 6th century BCE, developed a new way of understanding the world.

What was new? Before philosophy, the Greeks had a mythological view of nature. They saw the world around them as a family of personified natural forces. The earth was the mother who had given birth to all the others; the father was the sky; one of his brothers was the sea; another ruled the underground domain of the dead; beauty and wisdom was a daughter that grew out of the father's head, etc., etc. Natural events were understood as expressions of the whims and passions of these personified forces—the gods.

Miletus, it appears, played an important role in the development of the Greek writing system. Its script, which was adopted by Athens 403 BCE, became the basis of the contemporary Greek script. Writing, and the ability it gives one to contemplate fixed thought, plays an important role in the construction of concepts and the development of higher abstractions.

It appears that Thales was equipped with a new word—the cosmos. That word, which has been translated as the universe, or even nature, is about as abstract as one can get. It encompasses everything within the realm of experience—the Earth and everything on it, as well as the Sun, the Moon, the stars, the planets, and the space within which they exist. How can all of this diversity be captured by a single term?

Earlier, I argued that words—symbols—are critical for representing higher abstractions. A concept is a "one and a many." The word fruit is a "one"—an abstract term that encompasses "many" concrete things—apples, bananas, limes and watermelons. The word provides a "tool" that enables us to represent the fact that a number of different things share something fundamental in common. Apples, limes, etc., are all equally reproductive organs of plants. The development of words, then, goes hand in hand with the development of concepts.

What is it that makes it possible to include everything in experience within a single term? What do all the different things in the universe have in common? What is common to all of them, Thales argued, is that they are governed by the same principle.

What is that principle? There is no common principle that can be extracted from a mythological framework. There are just different actors with different agendas. Thales' great contribution was to remove the subjective dimension from the operation of the natural world and to see it as a physical mechanism.

The Greeks had not yet developed an adequate vocabulary for describing physical forces and processes, so Thales had to use something concrete to represent the physical qualities he was looking for. The principle of the cosmos, he proposed, is water. Water can take any form; it exists as a liquid, a solid, and a gas. Everything, he argued, arises from it, and everything returns to it.

That was a stroke of genius. The universe was seen as an integral physical entity—a One— governed by a single set of physical laws, by cause and effect rather than by the intentions of different gods. Based on this assumption, the early Greek philosophers were able to make truly amazing discoveries. They figured out that lunar eclipses were caused by the Earth coming between the Sun and the Moon, that the Earth was round, and even that its circumference was more or less 24,000 miles.

From the point of view of this book, what is important about this transition between mythology and philosophy is that it shows that while language may have opened the door to "objective" thinking, it did not automatically produce it.

Early humans were able to form cause and effect models of many of the concrete, practical problems they faced. Preliterate peoples learned immense amounts about the behavior of animals, the nutritional and medicinal value of different plants, what kind of materials made the best bows and tools, etc. However, they lacked the cognitive tools to make sense of the Big Picture—the larger physical and social/historical worlds they found themselves in.

By seeing everything as the product of physical cause and effect, the Greek philosophers developed the view that nature was logical, which, particularly in the hands of the Romans, became a powerful cognitive tool. While philosophical rationality never replaced religion or even, for that matter, what we would call superstition, it effected a merger with it. The dominant religion of the Romans was stoicism, a system of thought that could be traced back to the atomistic and largely materialistic philosophy of Epicurus.

The influence of Greek philosophy largely disappeared with the fall of Rome. While aspects of it were assimilated into Christianity (through Neo-Platonism), its scientific spirit would have to await the Enlightenment to be re-born.

In this book, I am arguing that humanity is on the cusp of a new cognitive revolution: one that involves a role for intentionality in the operation of nature; not, however, through the agency of gods, but of a humanity that understands its unique role and responsibilities within the natural order.

APPENDIX 5

I have used flower as an example of a term that represents something that cannot be represented at the level of imagery. On one level, it is not a good example. While the size, shape, and color of flowers may vary tremendously, nevertheless, most share a set of physical features that distinguish them from other parts of plants and that make them identifiable as a category: most have colorful petals and certain characteristic shapes. These common features reflect the fact that the shape and color of most flowers serve a common purpose—to attract insects and birds to disperse their pollen.

As a result of these shared physical features, people were able to recognize that different kinds of flowers were the same kind of thing. They used a single word to denote different kinds of flowers long before they understood the role of flowers in plant reproduction. Under these circumstances, the word for flower would not represent a concept; it would simply be a name—a label used to lump together objects that the senses are able to recognize as similar. A name is not a concept. A concept is not based on sensory information alone: it captures what the philosophers called the principle of something.

Before flowers could be understood as reproductive organs, people would have to understand how flowering plants reproduce. While people had long possessed a lot of practical information about plants, they did not know that they reproduced sexually. That was not discovered until 1694, when the German botanist Rudolf Camerarius published the results of his investigations.

Camerarius was the first person to clearly recognize that the different parts of flowers were functionally equivalent to the sex organs of animals. The anther was a male organ that produced pollen; pollen was functionally a form of semen; the pistil at the base of the flower in which the seeds developed was functionally an ovary; the pollen fused with and fertilized the seeds. A flower, then, is a structure that is organized to enable a plant to reproduce. This is a concept of a flower.

A concept is a One and a Many: the representation of something as a system or a process. It exists at a higher level of abstraction than perception. It was no simple matter for humans to recognize that plants, on one fundamental level, operated the same as animals. Camerarius' breakthrough came when he realized that the parts of flowers could be placed into a more general biological framework; that they were, in fact, male and female organs. That is when his perception of a flower was transformed into a concept—an abstract, logical model.

Language is the means by which humans are able to construct models of the world that exist at a higher level of abstraction than perception. While perceptual processes are always an important part of experience, it is conceptualization that carries human consciousness to another plane.

BIBLIOGRAPHY

Chapter 1

General Background for Discussion

Anthropology:

Benedict, Ruth. *Patterns of Culture.* Boston: Houghton Mifflin Company, 1959. (Orig. 1934.)
Boaz, Franz. *The Mind of Primitive Man*. New York: The Macmillan Company, 1938. (Orig. 1911.)
Kroeber, A.L. and Clyde Kluckhohn. *Culture: A Critical Review of Concepts and Definitions*. New York: Vintage Books, 1963. (Orig. 1952.)
Menand, Louis. "How Cultural Anthropologists Redefined Humanity." *The New Yorker*, August 26, 2019.
Radin, Paul. *Primitive Man as a Philosopher*. New York: D. Appleton Co., 1927.

Classical Analysis of Social Darwinism:

Hofstadter, Richard. *Social Darwinism in American Thought*. Boston: Beacon Press, 1955.

Sociobiology:

Maxwell, Mary. T*he Sociobiological Imagination*. Albany: State University of New York Press, 1991.

Wilson, E.O. Sociobiology: *The New Synthesis*. Cambridge, MA: Belknap Press, 1975.

Early Response of Cultural Anthropology to Sociobiology:

Sahlins, Marshall. *The Uses and Abuses of Biology*. Ann Arbor: The University of Michigan Press, 1977.

Chapter 2

deMenocal, Peter B. "Climate and Human Evolution." *Science*, Vol. 331, February 3, 2001.

Harmand, Sonia et al. "3.3-Million-Year-Old Stone Tools from Lomekwi, West Turkana, Kenya." *Nature*, Vol. 518: no. 7538, February 12, 2015.

Lovejoy, C. Owen. "Re-examining Human Origins in Light of *Ardipithecus ramidus*." *Science*, Vol. 326: issue 5949, October 2, 2009.

Lovejoy, C. Owen et al. "The Pelvis and Femur of *Ardipithecus ramidus*: The Emergence of Upright Walking." *Science*, Vol. 326, October 2, 2009.

__________________. "Careful Climbing in the Miocene." *Science*, Vol. 326, October 2, 2009. (See *Science*, Special Edition on *Ardipithecus*, October 2, 2009.)

Nowell, April and Iain Davidson. *Stone Tools and the Evolution of Human Cognition*. Boulder: University Press of Colorado, 2010.

Schick, Kathy and Nicholas Toth. *Making Silent Stones Speak: Human Evolution and the Dawn of Technology*. New York: Simon and Schuster, 1993.

Skinner, Matthew M. et al. "Human-like Hand Use in *Australopithecus aftricanis*." *Science*, Vol. 347, January 23, 2015.

Washburn, S. L. "Tools and Human Evolution." *Scientific American*, Vol. 203, 62-75, 1960.

Chapter 3

Kohler, Wolfgang. *The Mentality of Apes*. New York: Liveright Publishing, 1976.

Gibson, Kathleen Rita and Tim Ingold. *Tools, Language and Cognition in Human Evolution*. New York: Cambridge University Press, 1993.

Parker, Sue Taylor and Kathleen Rita Gibson. *Language and Intelligence in Monkeys and Apes*. New York: Cambridge University Press, 1994.

Visalberghi, Elisabetta and Loredana Trinca. "Tool Use in Capuchin Monkeys: Distinguishing Between Performing and

Understanding." *Primate*, Vol. 30: issue 4, October 1989.

Chapter 4

Hegel, G.W.F. *The Science of Logic.* (Translated by William Wallace.) New York: Oxford University Press, 1965.
Hume, David. *An Inquiry Concerning Human Understanding.* Indianapolis/Cambridge: Hackett Publishing Co., 1993.
Kant, Immanuel. *Immanuel Kant's Critique of Pure Reason.* New York: Macmillan & Co., 1963. (Also, St. Martin's Press.)

For an excellent discussion of cause and effect reasoning in chimpanzees and human children:

Premack, David and Ann. *Original Intelligence: Unlocking the Mystery of Who We Are.* New York: McGraw-Hill, 2003, chapter 8.

Chapter 5

MacNeilage, Peter F. et al. "Evolutionary Origins of Your Right and Left Brain." *Scientific American*, July 2009.
Damasio, Antonio. *Self Comes to Mind: Constructing the Conscious Brain.* New York: Pantheon Books, 2010.

Chapter 6

Call, Joseph and Michael Tomasello. "Does the Chimpanzee Have a Theory of Mind? Thirty Years Later." *Trends in Cognitive Sciences*, June 2008.
Donald, Merlin. *Origins of the Modern Mind.* Cambridge, MA: Harvard University Press, 1991.
Goldberg, Elkhonon. *The New Executive Brain: Frontal Lobes in a Complex World.* New York: Oxford University Press, 1991.
Gallop, Gordon Jr. "Chimpanzees: Self-Recognition." *Science*, Vol. 167: issue 3914, January 2, 1970.
Gallop, Gordon Jr. and James Anderson. "Mirror Self-recognition: A Review and Critique of Attempts to Promote and Engineer Self-recognition in Primates." *Primates*, Vol. 56, September 4,

2015.
Smaers, J.B. et al. "Primate Prefrontal Cortex Evolution: Human Brains are the Extreme of a Lateralized Ape Trend." *Brain Behavioral Evolution* (ePub), February 17, 2011.
Smaers, J.B. and C. Soligo. "Brain Reorganization, Not Relative Brain Size, Primarily Characterizes Anthropoid Brain Evolution." *Proceedings of the Royal Society B*, Vol. 280: issue 1759, May 22, 2013.

Recommended General Background

Donald, Merlin. *A Mind So Rare: The Evolution of Human Consciousness*. New York: W.W. Norton & Company, 2001.
Geary, David C. *The Origin of Mind: Evolution of Brain, Cognition, and General Intelligence*. Washington, DC: American Psychological Association, 2004.
Parker, Sue Taylor and Michael L. McKinney, editors. *The Origins of Intelligence: The Evolution of Cognitive Development in Monkeys, Apes, and Humans*. Baltimore: Johns Hopkins University Press, 1999.
Premack, David and Ann. *Original Intelligence: Unlocking the Mystery of Who We Are*. New York: McGraw-Hill, 2003.
Russon, Anne E. and David R. Begun, editors. *The Evolution of Thought: Evolutionary Origins of Great Ape Intelligence*. Cambridge: Cambridge University Press, 2004.
Tomasello, Michael. *A Natural History of Human Thinking*. Cambridge, MA: Harvard University Press, 2014.

Chapter 7

Bril, Blandine et al. "Functional Mastery of Percussive Technology in Nut-cracking and Stone-flaking Actions: Experimental Comparisons and Implications for the Evolution of the Human Brain." *The Royal Society*, Vol. 367: issue 1585, January 12, 2012.
Dimaggio, Erin N. et al. "Late Pliocene Fossiliferous Sedimentary Record and the Environmental Context of Early *Homo* from Afar, Ethiopia." *Science*, Vol. 347: issue 6288, March 20, 2015.

Drimmer, Frederick. *Captured by the Indians: 15 Firsthand Accounts, 1750-1870.* New York: Dover Publications, 1985.

Roach, Neil T. et al. "Elastic Energy Storage in the Shoulder and the Evolution of High-speed Throwing." *Nature* (Letter), Vol. 498, June 26, 2013.

Roach, Neil T. and Brian G. Richmond. "Clavicle Length, Throwing Performance and the Reconstruction of the *Homo erectus* Shoulder." *Journal of Human Evolution*, Vol. 80, March 2015.

Simpson, W.W. et al. "A Female *Homo erectus* Pelvis from Gona, Ethiopia." *Science*, Vol. 332, November 14, 2008.

Spoor, Fred et al. "Reconstructed *Homo habilis* Type OH7 Suggests Deep-rooted Species Diversity in Early *Homo*." *Nature*, Vol. 519, March 4, 2015.

Villmoare, Brian et al. "Early *Homo* at 2.8 Ma from Ledi-Geraru, Afar, Ethiopia." *Science*, Vol. 347: issue 6228, March 20, 2015.

Wrangham, Richard. *Catching Fire*. New York: Basic Books, 2009.

Chapter 8

Donald, Merlin. *A Mind So Rare*. New York: W.W. Norton & Company, 2001.

____________. "The Mimetic Origins of Language." in *The Oxford Handbook of Language Evolution*. Kathleen R. Gibson and Maggie Tallerman, editors. Oxford: Oxford University Press, 2013.

Gillespie-Lynch, Kristen Marie and Emily Sue Savage-Rumbaugh et al. "What Did Language Grow From? Ape Hands, Mouth, or Both." *Frontiers for Young Minds*, May 16, 2019.

Goodall, Jane. *The Chimpanzees of Gombe: Patterns of Behavior*. Cambridge, MA: Harvard University Press, 1986.

Mallery, Garrick. *Sign Language Among North American Indians*. Mineola, NY: Dover Publications, Inc., 2001.

Savage-Rumbaugh, Sue and Roger Lewin. *Kanzi: The Ape at the Brink of the Human Mind*. New York: John Wiley & Sons, Inc., 1994.

Seyforth, Robert M. and Dorothy L. Cheney. "Behavioral Mechanisms Underlying Vocal Communication in Nonhuman Primates." *Animal Learning and Behavior*, 1997.

Streidter, George F. *Principles of Brain Evolution*. New York: Oxford

University Press, 2005.
Terrace, Herbert S. *Why Chimpanzees Can't Learn Language and Only Humans Can.* New York: Cambridge University Press, 2019.
Tomasello, Michael. *Origins of Human Communication.* Cambridge, MA: MIT Press, 2008.

Chapter 9

Barton, Robert A. and Isabella Capellini. "Maternal Investment, Life, Histories, and the Costs of Brain Growth in Mammals." *Proceedings of the National Academy of Sciences,* Vol. 108: no. 15, April 12, 2011.
Fonseca-Azevedo, Karina and Suzana Herculano-Houzel. "Metabolic Constraint Imposes Tradeoff Between Body Size and Number of Brain Neurons in Human Evolution." *Proceedings of the National Academy of Sciences,* Vol. 109: no. 45, November 6, 2012.
Isler, Karen and Carel van Schaik. "The Expensive Brain: A Framework for Explaining Evolutionary Changes in Brain Size." *Journal of Human Evolution,* October 1, 2009. (2009B)
______________________. "Why Are There So Few Smart Mammals (but So Many Smart Birds)?" *Biology* (Letters), Vol. 5: no. 1, February 23, 2009. (2009A)
Kuzawa, Christopher et al. "Metabolic Costs and Evolutionary Implications of Human Brain Development." *Proceedings of the National Academy of Sciences,* September 9, 2014.
Leonard, Williams R. et al. "Fat Detection." Chapter 1 in *Taste, Texture and Post Ingestive Effects: Evolutionary Perspectives on Fat Ingestion and Metabolism in Humans.* Boca Raton: CRC Press/ Taylor and Francis, 2010.
Niven, Jeremy E. and Simon B. Laughlin. *Journal of Experimental Biology,* June 1, 2008. doi: 10.1242/jeb.017574.
Simpson, Scott W. et al. "A Female *Homo erectus* Pelvis from Gona, Ethiopia." *Science,* Vol. 322: no. 5904, November 13, 2008.
Tattersall, Ian. *Masters of the Planet.* New York: St. Martin's Press, 2012.
van Schaik, Carel. *Among Orangutans: Red Apes and the Rise of Human Culture.* Cambridge, MA: Belknap Press, 2004.

Chapter 11

Ambrose, Stanley H. "Paleolithic Technology and Human Evolution." *Science*, Vol. 291: no. 5509, March 2, 2001, 1748-1753.
Arbib, Michael. *How the Brain Got Language*. New York: Oxford University Press, 2012.
Callaway, Ellen. "Evidence Mounts for Interbreeding Bonanza in Ancient Human Species." *Nature* (News and Comment), February 17, 2016.
Culotte, Elizabeth and Ann Gibbons. "Almost All Living People Outside of Africa Trace Back to a Single Migration More Than 50,000 Years Ago." *Science*, September 21, 2016.
Gibbons, Ann. "World's Oldest *Homo sapiens* Fossils Found in Morocco." *Science* (online), June 7, 2017, (Archeology, Human Evolution) doi:10.1126/science.aan6934.
aan6934
Henshilwood, Christopher S. et al. "A 100,000-Year-Old Ochre-processing Workshop at Blombos Cave, South Africa." *Science*, Vol. 334: no. 6053, October 14, 2011.
Heinrich, Joseph. *The Secret of Our Success*. Princeton, NJ: Princeton University Press, 2016.
Hirst, K. Kris. "Omo Kibish (Ethiopia) — Oldest Known Example of Early Modern Humans." *ThoughtCo*. (https://www.thoughtco.com/omo-kibish-in-ethiopia-172040) October 1, 2018; updated Dec. 3, 2020.
Hublin, Jean-Jacques et al. "New Fossils from Jebel Irhoud, Morocco and the pan-African Origin of *Homo sapiens*." *Nature*, Vol. 546, June 8, 2017.
Jacobs, Zenobia and Richard G. Roberts. "Human History Written in Stone and Blood." *American Scientist*, Vol. 97: no. 4, July-August 2009.
Neubauer, Simon, Jean-Jacques Hublin, and Philipp Gunz. "The Evolution of Modern Human Brain Shape." *Science Advances*, Vol. 4: no. 1, January 2018.
Olson, Steve. *Mapping Human History: Genes, Race, and Our Common Origins*. Boston: Mariner Books, 2003.
Sanders, Robert. "160,000-Year-Old Fossilized Skulls Uncovered in Ethiopia are Oldest Anatomically Modern Humans."

University of California News, June 11, 2003.
Stringer, Chris. *Lone Survivors.* New York: St. Martin's Press, 2012.
Tattersall, Ian. *Masters of the Planet.* New York: St. Martin's Press, 2012.
__________. *Becoming Human: Evolution and Human Uniqueness.* San Diego/New York: Harcourt Brace & Company, 1998.
Wells, Spencer. *Deep Ancestry: Inside the Genographic Project.* Washington, DC: National Geographic, 2007.
Ziegler, Martin et al. "Development of Middle Stone Age Innovation Linked to Rapid Climate Change." *Nature Communications*, 4:1905, May 21, 2013.

Chapter 12

Arbib, Michael. *How the Brain Got Language.* New York: Oxford University Press, 2012.
Bickerton, Derek. *Language and Human Behavior.* Seattle: University of Washington Press, 1995.
Donald, Merlin. *A Mind So Rare.* New York: W.W. Norton & Company, 2001.
MacLarnon, Ann and Gwen Hewitt. "The Evolution of Human Speech: The Role of Enhanced Breathing Control." *American Journal of Physical Anthropology*, Vol. 109: no. 3, 341-63; July 1999.
Henrich, Joseph. *The Secret of Our Success.* Princeton, NJ: Princeton University Press, 2016.

Chapter 13

Bickerton, Derek. *Adam's Tongue: How Humans Made Language, How Language Made Humans.* New York: Hill and Wang, 2009.
__________. *Language and Human Behavior.* Seattle: University of Washington Press, 1995.
Burling, Robbins. *The Talking Ape: How Language Evolved.* New York: Oxford University Press, 2005.
Courchesne, Eric et al. "Neuron Number and Size in Prefrontal Cortex of Children with Autism." *Journal of American Medical Association*, Vol. 306: no. 18, November 1, 2011.
Grandin, Temple. *Thinking in Pictures: My Life with Autism.* New York:

Vintage Books, 2006.
_____________. *The Way I See It: A Personal Look at Autism and Asperger's*. Arlington, TX: Future Horizons, 2011.
_____________. *The Autistic Brain: Thinking Across the Spectrum*. New York: Houghton Mifflin Harcourt, 2013.
Hegel, G.W.F. *The Logic of Hegel.* Translated from *The Encyclopaedia of the Philosophical Sciences*, by Dr. William Wallace, M.A. LLD. London: Oxford University Press, 1965. (Orig. 1873.)
Keller, Helen. *The Story of My Life*. New York: W.W. Norton & Company, 2004.
Terrace, Herbert S. *Why Chimpanzees Can't Learn Language and Only Humans Can*. New York: Cambridge University Press, 2019.
Vygotsky, Lev. T*hought and Language*. Cambridge, MA: MIT Press, 1986.

Language Debate

Interesting debate over the biological vs. cultural basis of language, highlighting the problem of separating these two dimensions.

Chomsky argues that language (grammar) is the product of a specific biological "device"—the product of a random mutation. Everette argues that grammars are so varied that it is not possible for them to be based on a common biological mechanism. He argues that language is a cultural tool: that it is the product of an advanced human intelligence—not a specific, biological enabling "device." Both are right; and both are wrong.

Berwick, Robert C. and Noam Chomsky. *Why Only Us: Language and Evolution*. Cambridge, MA: MIT Press, 2016.
Everett, Daniel. *Language: The Cultural Tool.* New York: Vintage Books, 2013.
Hauser, Marc D., Noam Chomsky and W. Tecumseh Fitch. "The Faculty of Language: What Is It, Who Has It, and How Did It Evolve?" *Science*, Vol. 298, November 22, 2002.

See Pinker also. He sees language as biological, but as opposed to Chomsky, the product of natural selection.

Pinker, Steven. *The Language Instinct*. New York: William Morrow and Company, 1994.

Chapter 14

Bruner, Emilio et al. "Evidence for Expansion of the Precuneus in Human Evolution." *Brain Structure and Function*, Vol. 222: no. 2, 1053-1060, March 2017.
Cavanna, Andrea E. et al. "The Precuneus: A Review of its Functional Anatomy and Behavioral Correlates." *Brain*, Vol. 129: issue 3, March 1, 2006.
Montefinese, Maria. "Semantic Representation of Abstract and Concrete Words." *Journal of Neurophysiology*, doi:10:1152, 2019.
Neubauer, Simon et al. "The Evolution of Modern Human Brain Shape." *Science Advances,* January 24, 2018.

Pereira-Pedro, A.S. et al. "Midsagittal Brain Variation Among Nonhuman Primates: Insights into the Evolutionary Expansion of the Human Precuneus." *Brain, Behavior and Evolution*, Vol. 90: no. 3, 255-263, 2017.
Sahlins, Marshal. *Culture and Practical Reason*. Chicago: the University of Chicago Press, 1976.
Starowicz-Filip, Anna et al. "The Role of the Cerebellum in the Regulation of Language Functions." *Psychiatria Polska*, Vol. 51, no. 4, 661-671, 2017.
Wallentin, Michael et al. "Accessing the Mental Space — Special Working Memory Processes for Language and Vision Overlap in Precuneus." *Human Brain Mapping*, Vol. 29: issue 5, May 24, 2007.

Chapter 15

Bowels, Samuel and Herbert Gintis. *The Cooperative Species: Human Reciprocity and Its Evolution*. Princeton: Princeton University Press, 2011.
Dunbar, Robin. *Grooming Gossip, and the Evolution of Language*. Cambridge, MA: Harvard University Press, 1996.
Wilson, E.O. *The Social Conquest of the Earth*. New York: Liveright

Publishing/W.W. Norton & Co., 2012.
________. *On Human Nature*. Cambridge, MA: Harvard University Press, 1978.
Pfaff, Donald W. *The Neuroscience of Fair Play*. New York: Dana Press, 2007.
Presti, David E. *Foundational Concepts in Neuroscience*. New York: W.W. Norton & Company, 2016.
Lieberman, Matthew D. Social: *Why Our Brains are Wired to Connect.* New York: Crown Publishers, 2013.
Greene, Joshua. *Moral Tribes*. New York: Penguin Press, 2013.
Wexler, Bruce E. *Brain and Culture: Neurobiology, Ideology, and Social Change*. Cambridge, MA: MIT Press, 2008

Chapter 16

Michaels, David. *The Triumph of Doubt*. Oxford: Oxford University Press, 2020.

For documentation of previous mass extinctions produced by extreme climate events:

Brannen, Peter. *The Ends of the World*. New York: Harper Collins Publisher, 2017.
Ward, Peter and Joe Kirschvink. *A New History of Life*. New York: Bloomsbury Press, 2016.

www.ingramcontent.com/pod-product-compliance
Lightning Source LLC
LaVergne TN
LVHW050811061126
851857LV00012B/759
* 9 7 8 0 9 7 7 1 6 5 5 2 0 *